Contents

CONTENTS

OLIVER MORTON

Mapping Mars

SCIENCE, IMAGINATION, AND THE BIRTH OF A WORLD

PICADOR
NEW YORK

 For information, address Picador, 175 Fifth Avenue, New York, N.Y. 10010.

www.picadorusa.com

Picador® is a U.S. registered trademark and is used by St. Martin's Press under license from Pan Books Limited.

For information on Picador Reading Group Guides, as well as ordering, please contact the Trade Marketing department at St. Martin's Press.
Phone: 1-800-221-7945 extension 763
Fax: 212-677-7456
E-mail: trademarketing@stmartins.com

ISBN 0-312-24551-3 (hc)
ISBN 0-312-42261-X (pbk)

First published in Great Britain by Fourth Estate

First Picador Paperback Edition: September 2003

10 9 8 7 6 5 4 3 2 1

For Lieutenant Arthur Noel Morton, RNVR
Navigating officer, HMS *Hargood*, 1944–45
Lover of maps, lover of writing and loving father
—"Round about here, Sir."

"Are you going to move our stuff?"
"No, that's the view. We're in the picture."

—Exchange between William Fox
and Mark Klett in William L. Fox,
View Finder

THE WESTERN HEM

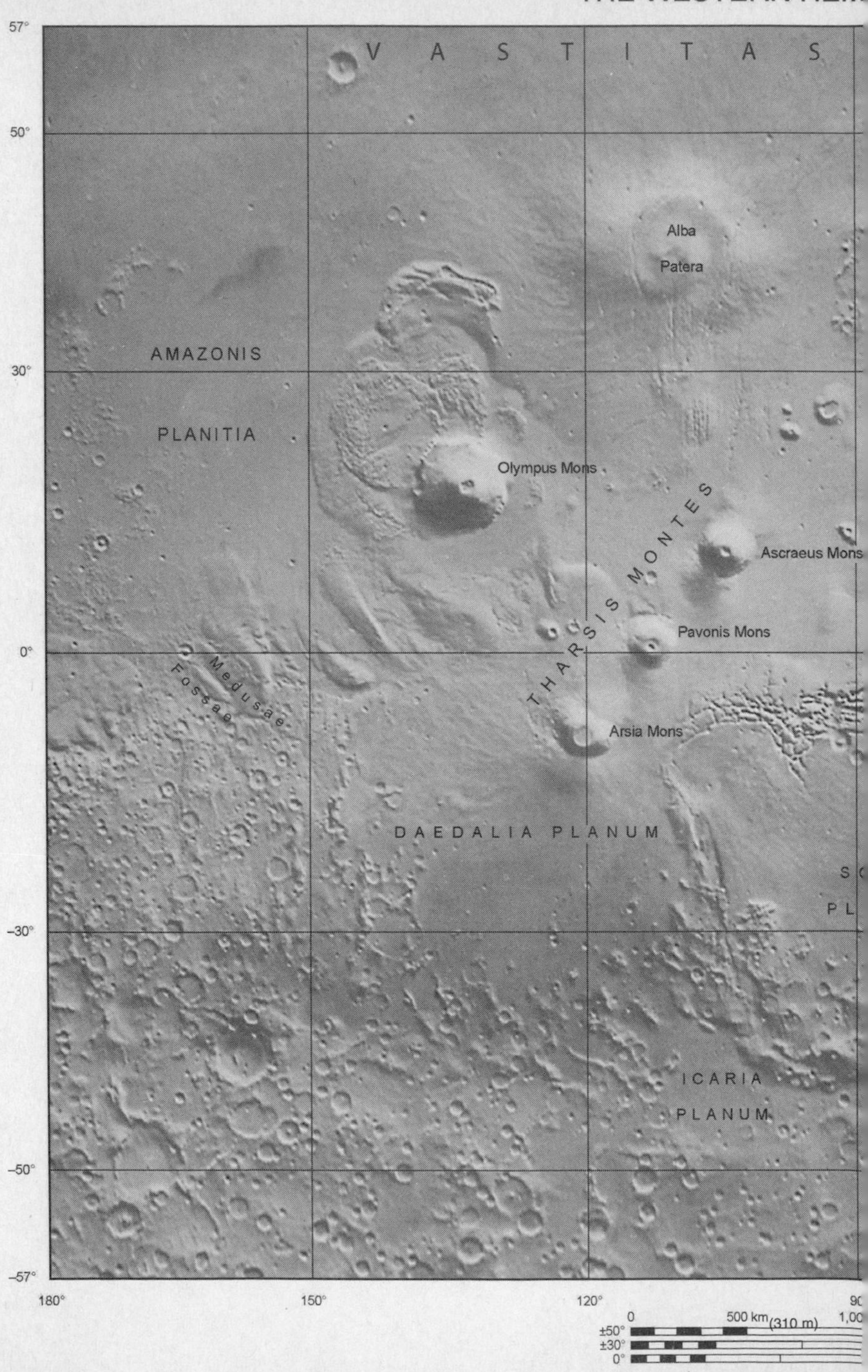

IISPHERE OF MARS
B O R E A L I S
ACIDALIA
PLANITIA
CHRYSE
PLANITIA
Kasei
Valles
Ares Vallis
TERRA
MERIDIANI
VALLES MARINERIS
LIS
NUM
NOACHIS
TERRA
ARGYRE
PLANITIA
Lowell
57°
50°
30°
0°
–30°
–50°
–57°
60°
30°
0°
(620 m)
2,000 km
(1240 m)

Introduction

There's a world on my wall.

Mountains, canyons, plains, and valleys, all a faded pinkish ochre, an even tone as plain as a color can be without being gray. The sun is to the west—shadows fall gently to the right. There are faults and rifts, ash flows and lava fields. There are creases and stretch marks, straight lines and strange curves. There are circles and circles and circles.

No cities. No seas. No forests and no battlegrounds. No prairies. No nations. No histories and no legends. No memories. Just features, features and names. Argyre and Hellas and Isidis. Olympus and Alba and Pavonis. Schiaparelli and Antoniadi, Kasei and Nirgal. Beautiful double-rimmed Lowell. Names from one world projected onto maps of another. Maps of Mars.

The maps on my wall, painstakingly painted about fifteen years ago, show the surface of Mars from pole to pole. They show volcanoes that dwarf their earthly cousins in age and size. They show the round scars of uncountable asteroid impacts, many far more violent than the one that killed off the Earth's dinosaurs. They show a canyon so long and deep it's as if the planet's tight skin has swollen and split. They show featureless plains and pockmarked ones, jumbled hummocky hills and strange creases that swarm together for thousands of miles, like the grain in a piece of timber. They show features perfectly earthlike and features so strange the Earth has no names for them. There's a world's worth of scientific puzzles here, some of them already tentatively answered, most still mysterious. There's a world's worth of possibilities. But there's no clear place to start the story.

If people had moved across the pinkish ochre—if they had grown vines on the terraces of Olympus, or herded goats through the Labyrinths of the Night; if legends haunted Tempe and the dales of Arcadia, or if in Ares Vallis ancient grudge had broken into new mutiny—then it would be easy. But there are none of those tales to tell. No gardens of Eden, no sacred springs, nowhere to start the story of a world.

Even stripped of people, with their cities and their borders and their histories, a map of Earth would not be this unyielding. Global truths and discrete units of geography would draw the eye. River catchments would tile the plains, mountain ranges would stand like the backbones of continents. There would be seas and islands, well defined. But Mars is not like that. It is continuous, seamless and sea-less. Its great mountains stand alone; there are no sweeping ranges, no Rockies or Alps or Andes. The rivers are long gone. There are no continents and there are no oceans, and thus there are no shores. Given patience, provisions, and a pressure suit you could walk from any point on the planet to any other. No edges guide the eye or frame the scene. Nowhere says: Start Here.

We might begin the story at one of the places that humanity has touched. In 1971 a Russian spacecraft crashed near Hellas, a vast basin in the southern hemisphere, while another landed more decorously on the other side of the planet, somewhere in or around the crater Ptolemaus. Two years later another Russian probe struck the surface somewhere near the dry valley called Samara. None sent back anything by way of a message. In 1976 America's more sophisticated *Viking* landers lowered themselves gently to sites in the northern plains of Chryse and Utopia, sending back panoramas of rock and rubble beneath pink-looking skies. But the *Viking*s eventually fell silent too, leaving Mars alone again. Preludes, not beginnings.

Twenty years later, the National Aeronautics and Space Administration's (NASA) little *Pathfinder,* cocooned in airbags, bounced to a halt in the rocky fields where Ares Vallis had once spewed out its floodwaters. It let loose *Sojourner*, the first of humanity's creations to travel on its own across the sands of Mars. That was a new beginning, the beginning of a grand age for Earth's robots. At the time of writing there have been automatic envoys sending data back from Mars ever since. But *Pathfinder*'s story cannot encompass the whole vast world in front of me. Not yet.

What about beginning on Earth? Some places here are very like locations there, perhaps close enough to be tied together by some sort of sympathetic story-magic. Maybe Antarctica, where the driest, coldest landscapes on Earth are regularly visited by scientists wanting to get some sense of a smaller, drier, colder world. Or Iceland, where permafrost and lava fight as once they did on Mars. Or the scablands of Washington State, ripped clean by floods like those that scoured *Pathfinder*'s landing site. Or Hawaii's volcanoes, near-perfect miniatures of the Martian giants. Or Arizona's Meteor Crater, where earthly geologists first came to grips with what a little bit of asteroid can do to the face of a planet, given enough speed. They are all places where one can learn about Mars, where the trained imagination can almost touch it. But none evokes the whole world.

We could cast our imaginations wider, to those who have tried to speak for all of Mars. To the astronomers looking at it with their telescopes, measuring all the qualities of light reflected from its surface, seeing seasons and imagining civilizations. Or to the writers inspired by those astronomical visions: H. G. Wells and Stanley Weinbaum, Arthur C. Clarke and Robert Heinlein, Ray Bradbury and Alexander Bogdanov and Edgar Rice Burroughs. Their imaginations took a point of light and turned it into a world of experience. But their Mars was never this one, the one that we only saw—that we could only ever see—after our envoys left the Earth and went there.

Only after our spacecraft reached its orbit could we see Mars for what it is, a planet with a surface area as great as that of the Earth's continents, all of it as measurable, as real as the stones in the pavement outside your door. After millennia of talking about worlds beyond our own, of heavens and hells and the Isles of the Hesperides, humanity now has such a world fixed in its sights, solid and sure. For the moment it is a world of science, untouchable but inspectable and oddly accessible, if only through the most complex of tools. But unlike the other worlds that scientists create with their imaginations and instruments—the worlds of molecular dynamics and of inflationary cosmology and all the rest of them—this one is on the edge of being a world in the oldest, truest, sense. A world of places and views, a world that would graze your knees if you fell on it, a world with winds and sunsets and the palest of moonlight. Almost a world like ours, except for the emptiness.

This book is about how ideas from our full and complex planet are projected onto the rocks of that simpler, empty one. The ideas discussed are mostly scientific, because it is the scientists who have thought hardest and best about the realities of Mars. It is the scientists who have fathomed the ages of its rocks, measured its resemblance to the Earth, searched for its missing waters, and—always—wondered about the life it might be home to. The stories they tell about the planet must have pride of place. But there are artists in here too, and writers, and poets, and people whose dreams take no such articulated form, but still focus themselves on the same rocks in the sky. They illuminate Mars; Mars illuminates them.

It's common to imagine that the human story on Mars will only start when humans actually get there, when they stand beneath its dusty sky and look around them at its oddly close horizon. I don't know who those people will be, or when they will get there, or where on the planet they will first set their feet. But I know that for all their importance, they will not be a new story's beginning, rather a new chapter. Their expectations and hopes are already being created on the Earth today, by the people in this book; the process of making Mars into a human world has already begun. And I know that their landing site is somewhere on the map in front of me, already charted, if not yet chosen.

Back to the maps, then; in particular to the 1:15,000,000 shaded-relief map of the surface published by the United States Geological Survey, its three sheets fixed to my office wall. It represents the planet as well as any single image could. But it's not just the representation of a planet. It's the embodiment of a process, a process that forged links between far-off Mars and the cartographers' drawing board point by point, feature by feature. It embodies links of reason and technology that ran through the cameras of now-dead spacecraft millions of miles away, and through the minds of the men who designed and controlled those cameras. Links that ran through empty space, carried by the faintest of radio waves, and through the great dishes that picked up those signals, and through the computers that wove them back into images. Links that ran through the eyes and minds and hands of the people who assembled the pictures produced by that great scientific adventure into a world they could see in their minds and draw on the paper in front of them, a world precise and publishable.

The maps themselves tell no single story. But the people who put those links together with technology and craft, mathematics and imagination—they have a story, one that lets the maps and the planet they are tied to come to life.

Where to begin to write about Mars? With the making of the maps.

Part I – Maps

"Now when I was a little chap I had a passion for maps. I would look for hours at South America, or Africa, or Australia, and lose myself in all the glories of exploration. At that time there were many blank spaces on the Earth, and when I saw one that looked particularly inviting on a map (but they all look that) I would put my finger on it and say, 'When I grow up I will go there.' The North Pole was one of these places, I remember. Well, I haven't been there yet, and shall not try now. The glamour's off. Other places were scattered about the Equator, and in every sort of latitude all over the two hemispheres. I have been in some of them, and . . . well, we won't talk about that. But there was one yet—the biggest, the most blank, so to speak—that I had a hankering after . . ."

—Marlow in Joseph Conrad's *Heart of Darkness*

Greenwich

> And then, as they sat looking at the ships and steamboats making their way to the sea with the tide that was running down, the lovely woman imagined all sorts of voyages for herself and Pa.
>
> —Charles Dickens, *Our Mutual Friend*

Maps of the Earth begin a short walk from the flat where I live. Go down the High Road, up Royal Hill toward the butcher's, left along Burney Street and then right onto Crooms Hill. At the corner, if you care for such things, you can see a blue plaque of the sort with which London marks houses where people who have made a significant contribution to human happiness once lived. In this case, it was the poet Cecil Day Lewis; as you climb the hill, you'll pass another one marking the home of Benjamin Waugh, founder of the National Society for the Prevention of Cruelty to Children.

Near the top of the hill sits a grand (but plaqueless) bow-fronted white house, called simply the White House. Walk around the White House's walled garden, down a little alleyway and through a gate in the high brick wall on your right, and you emerge into Greenwich Park. To your right, the beautiful semicircle of the rose garden; to your left a steep path lined by trees. And as you walk out onto the grass, London spread at your feet. As views go, it's not particularly extensive—the horizon is nowhere more than a dozen miles away and in many directions much closer—but it's vast in association. The once imperial cityscape is woven from threads that stretch throughout the world.

Across the river to the east sits the squat black-glass bulk of Reuters, information from around the globe splashing into its rooftop dishes. Upstream and on the near side sit the long, low workshops where for more than a century men have made undersea cables to tie the continents together. New skyscrapers devoted to global businesses sit in the redeveloped heart of the docks that used to handle the lion's share of the world's sea trade. Within the park itself there are plants from every continent except Antarctica. At its foot sits the old naval college, where generations of Britannia's officers, my late father included, learned to rule the waves.

Through it all the Thames runs softly, looping around the Isle of Dogs, a local feature leading, as Conrad says in *Heart of Darkness*, "to the uttermost ends of the Earth." Little sails down this umbilicus of empire now—but above it the new trade routes of the sky are sketched out by aircraft arriving and departing from London's four airports, carving their way through the air we all breathe and the stratosphere we shelter under. To the west the Thames beneath them is still daytime blue; to the east it is already evening dark.

Dawn may feel like an intervention by the sun, rising above a stationary Earth; sunset reveals the truth of the Earth's turning, a slipping away into night. That turning defines two unique, unmoving points on the surface of the Earth: the poles, the extremes of latitude. Add one more point—just one—and you have a coordinate system that can describe the whole world, a basis for all the maps and charts the sailors and pilots need, a way of deciding when days start and end. And that third point is right in front of you, the strongest of all Greenwich's links to the rest of the Earth. In the middle of the park is the old Royal Observatory, a little gathering of domes perched clubbily on a ridge. Within the observatory sits a massive metal construction called a transit circle. The line passing through the poles and through that transit circle is the Earth's prime meridian: 0 degrees, 0 minutes, 0 seconds. All earthly longitudes are measured with respect to that line through Greenwich Park.

The English have taken the Greenwich meridian as the starting point for longitudes since the observatory was founded in the seventeenth century. But it wasn't until the late nineteenth century—at a time when its home in Greenwich was under the stewardship of Sir George Airy, Astronomer Royal, the man who had that great transit circle built—that the Greenwich meridian was formally adopted by the rest of the world. With worldwide navigation a commonplace, and with telecommunications making almost instantaneous contact between continents a possibility, there was a need for a single set of coordinates to define the world's places and time zones. Over the years a variety of possible markers to define this prime meridian were suggested—islands, mountains, artifacts like the Great Pyramid or the Temple in Jerusalem. But a meridian defined by an observatory seemed best. In 1884, at a conference in Washington, D.C., and over spirited French opposition, Greenwich was chosen. Airy's transit circle came to define the world.

Airy was, by all accounts, an uninspiring but meticulous man. He recorded his every thought and expenditure from the day he went up to Cambridge University to more or less the day he died, throwing no note away, delighting in doing his own double-entry bookkeeping. He applied a similar thoroughness to his stewardship over the Royal Greenwich Observatory, bringing to its workings little interest in theory or discovery but a profound concern for order, which meant that the production of tables for the Admiralty (the core of the observatory's job) was accomplished with mechanical accuracy. He looked at the heavens and the Earth with precision, not wonder, and though he had his fancies, they were fancies in a similar vein—ecstasies of exactitude such as calculating the date of the Roman invasion of Britain from Caesar's account of the timing of the tides, or meticulously celebrating the geographical accuracy of Sir Walter Scott's poem "The Lady of the Lake." This was a man whose love of a world where everything was in its place would lead him to devote his own time to sticking labels saying "empty" on empty boxes rather than disturb the smooth efficiency of the observatory by taking an underling from his allotted labors to do so for him. After more than forty-five years of such service Airy eventually retired two hundred yards across the park to the White House on Crooms Hill, where he died a decade later.

It's a little sad that the White House doesn't carry a blue circular plaque to commemorate Airy's part in the happiness brought to humanity by a single agreed-upon meridian, but surely there are monuments elsewhere. Maybe Ipswich has an Airy Street; he grew up there and remained fond of the place, arranging for his great transit circle to be made at an Ipswich workshop. There must be a bust of him in the Royal Astronomical Society or a portrait in some Cambridge common room. And even if there are none of these things, there is something far grander. Wherever else astronomers go when they die, those who have shown even the faintest interest in the place are welcomed onto the planet Mars, at least in name. By international agreement, craters on Mars are named after people who have studied the planet or evoked it in their creative work—which mostly makes Mars a mausoleum for astronomers, with a few science fiction writers thrown in for spice. In the decades since the craters of Mars were first discovered by space probes, hundreds of astronomers have been thus immortalized. But none of them has a crater more fitting than Airy's.

A Point of Warlike Light

"I've never been to Mars, but I imagine it to be quite lovely."

—Cosmo Kramer in *Seinfeld*
("The Pilot (1)," written by Larry David)

Mars had an internationally agreed prime meridian before the Earth did. In 1830 the German astronomers Wilhelm Beer and Johann von Mädler, famous now mostly for their maps of the moon, turned their telescope in Berlin's Tiergarten to Mars. The planet had been observed before. Its polar caps were known, and so was its changeability; the face of Mars varies from minute to minute, due to the Earth's distorting atmosphere, and from season to season, due to quite different atmospheric effects on Mars itself. There are, though, some features that can be counted on to stick around from minute to minute and season to season, the most notable being the dark region now called Syrtis Major, then known as the Hourglass Sea. To calculate the length of the Martian day, Mädler (Beer owned the telescope—Mädler did most of the work) chose another, smaller dark region, precisely timing its reappearance night after night. He got a figure of 24 hours 37 minutes and 9.9 seconds, 12.76 seconds less than the currently accepted figure. That this length of time is so similar to the length of an earthly day is complete coincidence, one of three coincidental similarities between the Earth and Mars. The second coincidence is that the obliquity of Mars—the angle that its axis of rotation makes with a notional line perpendicular to the plane of its orbit—is, at 25.2°, very similar to the obliquity of the Earth. The third is that though Mars is considerably smaller than the Earth—a little more than half its radius, a little more than a tenth its mass—its surface area, at roughly a third of the Earth's, is quite similar to that of the Earth's continents.

When Mädler came to compile his observations into a chart in 1840, mathematically transforming his sketches of the disc of Mars into a rectangular Mercator projection, he declined to name the features he recorded, but did single out the small dark region he had used to time the Martian day as the site of his prime meridian, centering his map on it. Future astronomers followed him in the matter of the meridian while eagerly making good his oversight in the matter of names. Father Angelo Secchi, a Jesuit at the Vatican observatory, turned the light and dark patches into continents and seas, respectively, as astronomers had done for the moon, and gave the resulting geographic features the names of famous explorers—save for the Hourglass Sea, which he renamed the "Atlantic Canale," seeing it as a division between Mars's old world and its new. In 1867 Richard Proctor, an Englishman who wrote popular astronomy books, produced a nomenclature based on astronomers, rather than explorers, and gave astronomers associated with Mars pride of place. His map has a Mädler Land and a Beer Sea, along with a Secchi Continent. Observations made by the Astronomer Royal in the 1840s—he was interested in making more precise measurements of the planet's diameter—were commemorated by the Airy Sea. Pride of place went to the Rev. William Rutter Dawes, a Mars observer of ferociously keen eyesight, perceiving, for example, that the dark patch Mädler had used to mark the prime meridian had two prongs. (Dawes's far-field acuity was allegedly compensated by a visual deficit closer to home; it is said he could pass his wife in the street without recognizing her.) So great was Dawes's influence on Proctor—or so small was the number of astronomers associated with Mars—that his name was given not just to the biggest ocean but also to a Continent, a Sea, a Strait, an Isle and, marking the meridian, his very own Forked Bay.

Proctor's names had two drawbacks, one immediately obvious, one revealed a decade later. The obvious drawback was that an unhealthy number of the people commemorated on Mars were now British. When the French astronomer Camille Flammarion revised Proctor's nomenclature for his own map of 1876, various continentals—Kepler, Tycho, Galileo—were given grander markings. One continental on whom Proctor had looked with favor, though, was thrown off: Perhaps influenced by the Franco-Prussian war, Flammarion resisted having the most prominent dark patch on the planet called the Kaiser Sea, even if Proctor had named it such in honor of Frederik Kaiser of the Leiden Observatory. The Hourglass Sea became an hourglass again, though this time in French: Mer du Sablier.

Proctor's other problem was more fundamental. The features he had marked on his map, whatever their names, did not match what other people saw through their telescopes. In 1877, Mars was in the best possible position for observation; it was at its nearest to the sun (a situation called perihelion) and at its nearest to the Earth (a situation called opposition), just thirty-five million miles away. Impressive new telescopes all over the world were turned to Mars and revealed its features in more detail than ever before. The maps based on observations made that year were almost all better than Proctor's; and the map made by Giovanni Schiaparelli, a Milanese astronomer, on the basis of these observations, provided a new nomenclature that overturned all others.

Schiaparelli was not interested in celebrating his peers and forebears; he wanted to give Mars the high cultural tone of the classics. In the words of Percival Lowell, an American astronomer who was to make Mars his life work, it was an "at once appropriate and beautiful scheme, in which Clio [muse of poetry and history] does ancillary duty to Urania [muse of astronomy]." To the west were the lands beyond the pillars of Hercules, such as Tharsis, an Iberian source of silver mentioned by Herodotus, and Elysium, the home of the blessed at the far end of the Earth. Beneath them, part of the complex dark girdle strung around Mars below its equator, were the sea of sirens, Mare Sirenum, and Mare Cimmerium, the sea that Homer put next to Hades, "wrapped in mist and cloud." Then we come to the Mediterranean regions: the Tyrrhenian Sea and the Gulf of Sidra (Syrtis Major, the long-observed hourglass) dividing bright Hellas and Arabia. Along the far side of Arabia sits the Sinus Sabeus, a gulf on the fragrant coast of Araby, home to the Queen of Sheba. Beyond Arabia begins the Orient, with Margaritifer Sinus, the bay of pearls on the southern coast of India, and the striking bright lands of Argyre (Burma) and Chryse (Thailand). Finally, in the dark region others had called the eye of Mars, Schiaparelli placed Solis Lacus, the lake of the sun, from which all dawns begin.

Do not think for a moment that this means a good classical education will help you find your way around Mars. For a start, due to the way telescopes invert images, everything is flipped around: Greece is south of Libya, Burma west of Arabia. What's more, Schiaparelli's geography was often more allusive than topographical. His planet is 360° of free association. Thus Solis Lacus is surrounded by areas named for others associated with the sun: Phoenix, Daedalus, and Icarus. The sea of the sirens borders on the sea of the muses, presumably because Schiaparelli wanted to provide opportunity for their earthly feud to continue. Elysium leads to Utopia.

For the most part he did not explain his nominal reasoning very exactly, but there are exceptions, most notably right in the middle of the map, at the point where dark Sinus Sabeus gives way to Sinus Margaritifer, somewhere between Arabia and the Indies, a place he called Fastigium Aryn. "As Mädler," Schiaparelli wrote, "I have taken the zero-point of the areographic longitudes there, and following this idea I have given it the name of Aryn-peak or Aryn-dome, an imaginary point in the Arabian sea—which was long assumed by the Arabic geographers and astronomers as the origin of the terrestrian longitudes."

By the time he was through with Mars, Schiaparelli had given 304 names to features on its surface and though there was a Proctorite resistance—"'Dawes' Forked Bay it will ever be to me, and I trust to all who respect his memory," wrote Nathaniel Green, who painted a lovely map of Mars after observing the planet from Madeira during the opposition of 1877—it foundered. Schiaparelli's proper names were triumphant and have in large part lasted until today. It was his common nouns that caused the problems. Schiaparelli saw a large number of linear features on the face of the planet and called them "canali"—channels. Schiaparelli claimed to be agnostic as to the nature of these channels—they might have been natural, or they might have been artificial. Percival Lowell, his most famous disciple, plumped firmly for the artificial interpretation.

Lowell's reasoning went like this. Mars is habitable, but its aridity makes the habitability marginal; if there were intelligences on Mars, they would do something about this; the obvious thing to do would be to build a network of long, straight canals. And since this is what we see when we look at Mars, this is what must have happened.

With this leap of the imagination, Lowell created one of the most enduring tropes of science fiction: Mars as a dying planet. It would live on in the works of H. G. Wells, Edgar Rice Burroughs, Leigh Brackett, and many, many others. And if his interpretation of what he saw did not win as much favor among his astronomical colleagues as it did in the popular imagination, it was not because the idea of life on Mars seemed too far-fetched. Observations of the way the planet's brightness and color seemed to change with the seasons made plant life there seem almost certain; if plants, why not animals and why not intelligence? The weightiest argument against Lowell's Martians was simply that, over time, other better observers consistently failed to see the canals as continuous and linear, if they saw them at all. The lack of evidence of engineering, not the implausibility of life on Mars, was what counted against Lowell—a belief in life on Mars was quite commonplace.

Today this easy acceptance seems rather remarkable. At the beginning of the twenty-first century, when the possibility of life elsewhere has become the central preoccupation of space exploration, its discovery is routinely held up as the most important discovery that could ever be made. What accounts for this change?

A large part of the answer lies in the nature of astronomy. Copernicus's proposal that the Earth was not the center of the solar system changed the way that astronomers looked at the sky. If the Earth was no longer the fixed center, then it was a wandering star like the five that shuffled back and forth across the zodiac: a planet. Previously unique, now it was one member of a class and must have similarities to its classmates. The world had become a planet and so the planets must become worlds, a process accelerated by the Galilean discovery that, like the Earth, the planets were round and had features. In this context it was quite normal to believe that one of the things that the planets had in common was life, especially since, after Copernicus, many astronomers tended to go out of their way to deny the Earth any special attributes. As Lowell put it in

Mars (1896), "That we are the only part of the cosmos possessing what we are pleased to call mind is so Earth-centered a supposition, that it recalls the other Earth-centered view once so devoutly held, that our little globe was the point about which the whole company of heaven was good enough to turn. Indeed, there was much more reason to think that then, than to think this now, for there was at least the appearance of turning, whereas there is no indication that we are sole denizens of all we survey, and every inference we are not." A Copernican stance could easily lead astronomers to the assumption of life, not lifelessness, as the status quo.

Another part of the answer is that in Lowell's day a belief in life on Mars was largely without consequences. As Alfred Lord Tennyson noted as early as 1886, our astronomical observations of planets and our dreams of what might transpire on them were separated by a vast gulf:

> Hesper—Venus—were we native
> to that splendor or in Mars,
> We should see the Globe we groan in,
> fairest of their evening stars.
>
> Could we dream of wars and carnage,
> craft and madness, lust and spite,
> Roaring London, raving Paris,
> in that point of peaceful light?

Life on Mars might be likely, it might be inevitable, it might even be intelligent, but the possibility of people ever actually visiting Mars—or Martians visiting Earth—was more or less pure fancy. This made Martians fascinating but not important, rather in the way of dinosaurs—another turn-of-the-century craze. Whatever evidence scientists might find of dinosaurs, or speculations they might produce about them, without a time machine encounters with dinosaurs were impossible. Similarly, without a space machine, encounters with Martians were impossible.

So while there might be intelligent Martians, there could be no links of history or interest between them and us. This gave the Martians an interesting rhetorical niche that they quickly made their own: "The man from Mars" became the quintessential intelligent outsider, unswayed by any relevant prior worldliness, unattached to custom. He retains that position to this very day; his natural habitat is the newspaper op-ed page and other didactic or satirical environments, but he turns up elsewhere, too. Temple Grandin, the highly articulate woman with autism in Oliver Sacks's *An Anthropologist on Mars,* applies the titular image to herself as a way of stressing her disassociation from the ways of the world around her; the wonderfully innocent yet artfully contrived metaphors of the poems in Craig Raine's *A Martian Sends a Postcard Home* led to a whole school of poetry (if a small one) being dubbed "Martianism." One of the most influential science fiction novels of the twentieth century, Robert Heinlein's *Stranger in a Strange Land,* achieves its impact by showing us the Earth through the eyes of a true "man from Mars"—a human brought up on Mars by Martians.

Rhetorical devices aside, believing in Martians made little difference to the earthly lives of Lowell's readers and this, I suspect, is one of the things that made them easy to believe in. Another spur to belief was the difference that the existence of Martian minds made to the way earthly imaginations saw Mars. One of the Copernican ways in which Martians made the planet Mars a world like the Earth was that they made it a place experienced from the inside, a site for subjectivity. Without minds, Lowell argued, Mars and the other planets were "mere masses of matter"—places without purpose, frightening voids. With minds, they were worlds.

To Lowell, there was no really useful or involving way to think about a planet except as a world inhabited and experienced by mind. The space age, though, has brought us new ways of seeing beyond the Earth and changed our way of thinking about what we see. Our spacecraft, tools of observation but hardly observers in themselves, have shown us things we know cannot be witnessed directly or experienced subjectively, but which can still fascinate. The post-Copernican elision between worlds (structures of shared experience and history) and planets (vast lumps of rock and metal and gas that orbit a fire yet vaster) has been rewritten. Yes, the Earth that is our world is also a planet. But not all planets are worlds. We no longer need the point of view of a mythical Martian to imagine Mars, or to convince us that Mars might be worth imagining. Now that our spacecraft have been there we can know it intimately from the outside, know it as an objective body rather than a subjective experience. We can measure and map its elemental composition and its wind patterns and its topography and its atmospheric chemistry and its surface mineralogy. The planet Mars can fascinate us just for what it is.

If the space age has opened new ways of seeing mere matter, though, it has also fostered a strange return to something reminiscent of the pre-Copernican universe. The life that Lowell and his like expected elsewhere has not appeared, and so the Earth has become unique again. The now-iconic image of a blue-white planet floating in space, or hanging over the deadly deserts of the moon, reinforces the Earth's isolation and specialness. And it is this exceptionalism that drives the current scientific thirst for finding life elsewhere, for finding a cosmic mainstream of animation, even civilization, in which the Earth can take its place. It is both wonderful and unsettling to live on a planet that is unique.

Yet if the Earth is a single isolated planet, the human world is less constrained. The breakdown of the equation between planets and worlds works both ways. If there can now be planets that are not worlds, then there can be worlds that spread beyond planets—and ours is doing so. Our spacecraft and our imaginations are expanding our world. This projection of our world beyond the Earth is for the most part a very tenuous sort of affair. It is mostly a matter of imagery and fantasy. Mars, though, might make it real—which is why Mars matters.

Mars is not an independent world, held together by the memories and meanings of its own inhabitants. But nor is it no world at all. More than any other planet we have seen, Mars is like the Earth. It's not very like the Earth. Its gravity is weak, its atmosphere thin, its surface sealess, its soil poisonous, its sunlight deadly in its levels of ultraviolet, its climate beyond frigid. It would kill you in an instant. But it is earthlike enough that it is possible to imagine some of us going there and experiencing this new part of our human world in the way we've always experienced the old part—from the inside.

The fact that humans could feasibly become Martians is the strongest of the links between Mars and the Earth.

At the beginning of the space age—at the moment when it became clear to all that Mars might indeed one day be experienced subjectively—the International Astronomical Union stepped in to clean up the planet's increasingly baroque nomenclature. Thanks to the efforts of Schiaparelli, Lowell, and Eugène Michael Antoniadi, whose beautifully drawn charts had become the standard, the planet had come to boast 558 names for an uncertain number of features. In 1958 the IAU experts settled on 128 named regions and features, with 105 of the names coming from Schiaparelli. Then the first spacecraft images came back and the stalwarts of the IAU needed not only more names but also new rules by which names could be assigned. It was at this point that the convention of naming craters for people with an interest in the planet was laid down. Proctor's astronomical pantheon was reconvened—Dawes, Secchi, Mädler, Beer, and the rest of them all got craters, as did Proctor himself.

And in 1972 the International Astronomical Union established for all time the precise location of the Martian meridian. Lacking a transit circle made of good Ipswich steel—or, for that matter, any ancient monuments—the IAU's working group had to use a natural landmark for their zero. They chose the geometrical center of a small, nicely rounded crater in the middle of a larger crater thirty-five miles across. They called that larger crater Airy.

Mert Davies's Net

There is a passage in the oeuvre of William F. Buckley Jr. in which he remarks that no writer in the history of the world has ever successfully made clear to the layman the principles of celestial navigation. Then Buckley announces that celestial navigation is dead simple, and that he will pause in the development of his narrative to redress forever the failure of the literary class to elucidate this abecederian technology. There and then—and with awesome, intrepid courage—he begins his explication: and before he is through, the oceans are in orbit, their barren shoals are bright with shipwrecked stars.

—John McPhee, *In Suspect Terrain*

It was Merton Davies who put Airy in his prime position. Mert is a kindly man, tall and thin, dignified but rather jolly. Everyone who knows him speaks fondly of him. You might imagine him embodying decent reliability in a Frank Capra film even before learning that he has worked for the same outfit more than fifty years. But it's hardly been a small-town life. Mert Davies was one of the pioneers of spy satellites, one of the small cadre of technical experts who changed the facts of geopolitical life by letting cold warriors see the world over which they were at war from a totally new perspective. After that, he became one of only two people to have played an active role in missions to every planet save Pluto.* He has reshaped—quite literally—the way that earthlings see their neighbors in space. Davies is the man chiefly responsible for the "control nets" of most of the solar system's planets and moons—complex mathematical corsets that hold the scientific representations of those planetary surfaces together.

*The other, according to Caltech (California Institute of Technology) professor Bruce Murray, is Murray's Caltech colleague Ed Danielsen.

The first control net that he created served as the basis for the first maps of Mars made using data from spacecraft, rather than observations from Earth. Compiled from fifty-seven pictures sent back when *Mariner 6* and *Mariner 7* flew past the planet in 1969, that first net tied together 115 points. When I met Davies in his office in Santa Monica thirty years later, his latest Martian control net held 36,397 points from 6,320 images. Well into his eighties, Davies was still hard at work augmenting it further.

Davies had been interested in astronomy since boyhood, an interest he had shared with those close to him. In 1942, when he was working for the Douglas Aircraft corporation in El Segundo, California, he started courting a girl named Louise Darling. His interests made their dates a little unusual. Davies had started making a twelve-inch telescope, a demanding project. "I had a hard time finishing it," he recalls. "The amount of grinding it took and the difficulty of polishing that big a surface was a little bit over my head. I would take her with me to polish." And so she entered the world of grinding powder and the Foucault test, a simple but wonderfully precise way of gauging a mirror's shape, which allows an amateur with simple equipment to detect imperfections as small as two millionths of an inch. Unorthodox courtship, but it worked. When I met Mert in 1999 he and Louise had been married for more than fifty years.

Just after the war, Davies heard that a think tank within Douglas was working on a paper for the Air Force about the possible uses of an artificial satellite. He applied to join the team more or less on the spot. The think tank soon became independent from Douglas and, as the RAND Corporation, it went on to play a major role in defining America's national-security technologies and strategies throughout the Cold War. In the early 1950s Davies and his colleagues looked at ways to use television cameras in space in order to send back images of the Soviet Union. Then they developed the idea of using film instead of television—experience with spy cameras on balloons showed that the picture quality could be phenomenal—and returning the exposed frames to Earth in little canisters. The idea grew into the Corona project, which, after a seemingly endless run of technical glitches and launch failures at the end of the 1950s, became a spectacularly successful spy-satellite program.

While Corona was in its infancy, Davies was seconded to Air Force intelligence at the Pentagon, where he used the new American space technology to try to figure out what Russian space technology might be capable of. When he returned to Santa Monica in 1962, he was ready for a change. Spy satellites were no longer exciting future possibilities for think-tank dreamers, but practical programs controlled by staff officers and their industry contractors. And there was another problem. "A lot of the work at RAND was going into Vietnam—my colleagues were working on reconnaissance issues there—and I wanted no part of that."

Happily, an alternative offered itself in the form of Bruce Murray, an energetic young professor from the California Institute of Technology (Caltech) in Pasadena on the other side of Los Angeles. Murray was an Earth scientist, not an astronomer. His first glimpse of Mars through a telescope wasn't a childhood epiphany in the backyard. It was a piece of professional work from the Mount Wilson Observatory. Late as it was, though, that first sight provided emotional confirmation for Murray's earlier intellectual decision that the other planets were something worth devoting a lifetime's study to. When Murray looked at Mars through the world-famous sixty-inch telescope, he was not just seeing an evocative light in the sky; he was seeing a world's worth of new geology, a planet-sized puzzle that he and his Caltech colleagues were determined to crack. Their tool was to be the Jet Propulsion Laboratory (JPL), a facility that Caltech managed on behalf of the federal government. JPL, in the foothills of the San Gabriel Mountains, had been a center for military aerospace research since the war. In 1958 the Army ceded it to the newly founded National Aeronautics and Space Administration (NASA), as part of which it would become America's main center for planetary exploration. By 1961, JPL was planning NASA's first Mars mission, *Mariner 4.* The man in charge of building a camera for it was Robert Leighton, a Caltech physics professor. He asked a geologist he knew on the faculty, Bob Sharp, to help him figure out what the camera might be looking at. Sharp asked his eager young colleague Murray to join the team.

Murray and Davies met in 1963; with three young children to support, Murray was keen for some extra income and so found consulting for RAND congenial. He and Davies quite quickly became close friends and Mert started to think he might want to get involved in Murray's end of the space program. After all, he had the right credentials: He had been in the space business since the days of the V2 and he had some experience in interpreting images of both the Earth and moon as seen from orbit. (At the Pentagon he had analyzed Russian pictures of the far side of the moon to see whether they might be fakes.) When *Mariner 4*'s television camera sent back its image-data—a string of twenty-one grainy pictures covering just 1 percent of the planet's surface—Davies was as surprised as almost everybody else to see that it looked not like an earthly desert but like the pockmarked face of the moon, or the aftermath of a terrible war. The space program was important (Murray and his colleagues would brief the president) but it was also open (they briefed him in front of the cameras). Out among the planets there was no risk of finding yourself in a conflict you wanted no part of, or of having to keep work secret from all but your closest colleagues.

By the time *Mariner 6* and *Mariner 7* were sent to Mars four years later, in 1969, Davies was a key part of the team dealing with the images they sent back. His particular contribution was to work on the mathematical techniques needed to turn the disparate images into the most reliable possible representation of the planet.

Since the seventeenth century, when Willebrord Snell of Leiden first refined the procedure into something like its modern form, earthbound mapmakers have turned what can be seen into what can be precisely represented through surveying. Decide on a set of landmarks—Snell and his countrymen liked churches—and then, from each of these landmarks, take the bearings of the other landmarks nearby. From this survey data you can build up a network of fixed points all across the landscape. Plot every point on your map according to measurements made with respect to things in this well-defined network and it will be highly accurate. If, unlike Holland, your country is large, mountainous, and only sparsely supplied with steeples, setting up a reliable network in the first place can be hard work—the United States wasn't properly covered by a single mapping network until the 1930s, when abundant Works Progress Administration labor was available to help with the surveying. But the principle of measuring the angles between lines joining landmarks has been used in basically the same way all over the Earth.

Two problems make the mapping of other planets different, one conceptual, one practical. On Earth, experience allows you to know what the features you are mapping are: Hills, valleys, forests, and so on are easily recognized for what they are. While the pictures a spacecraft's cameras send back may be very good, this level of understanding is just not immediately available. When the first images of Mars were sent back by *Mariner 4* they were initially unintelligible to Murray and the rest of the imaging team. Before the researchers even started on a physical map, they needed a conceptual one, a way of categorizing what was before their eyes. How to do this—how to see what had never been seen before—was the besetting problem of early planetary exploration.

The practical problem is that unlike an earthly surveyor, you can't wander around the surface of an alien planet making measurements at leisure. Your only viewpoint is that of a spacecraft flying past the surface at considerable distance and speed. So you not only don't know what you're looking at; you're also none too sure of where you're looking from. A spacecraft's position is not a given, like that of a church. It is something that its controllers have to continuously work out. What they know for sure is how fast it is receding from the Earth, because that causes frequency changes in its radio signals. To find out where the spacecraft actually is, this information is compared with estimates of where the spacecraft thinks it is—the primary tools here are small onboard cameras called star trackers—and calculations of where it ought to be, derived from measurements and models of all the forces—the gravity of the sun and the planets, the gentle nudges from onboard thrusters—that are shaping its trajectory. If all is going well, the calculations based on all these observations fall into line to produce a consistent picture.* Although this may be accurate enough for navigation, it is not accurate enough for mapmaking. You don't know precisely where the spacecraft is, or precisely which way its camera is pointing, or, for that matter, precisely where the surface of the planet is. So you can't say exactly what bit of the planet you're looking at in any given picture.

Working around these problems involved Davies in a huge amount of laborious cross-checking and number crunching (Airy himself would have loved it, I suspect). First he had to put together a set of clearly distinguishable features that appeared in more than one of the pictures—the centers of craters, for the most part. The precise locations of these features within the individual frames in the data sent back by the spacecraft then had to be put into a set of mathematical equations along with the best available figures for the spacecraft's position when each picture was taken, and the direction in which the camera was pointing at the time. Then he had to add in factors describing the distortions the cameras were known to inflict on the pictures they took. Once all this was done, the whole calculation had to be fed into a computer on punch cards; the computer then ground through possible solutions until it came to one that made the values of all the variables in the equations consistent. Those values defined a specific way of arranging the set of surface features in three dimensions—imagine it as a framework of dots linked by straight lines—which came as close as possible to satisfying all the data. Effectively, the final answer said, "If the reference points you've specified are arranged in just this way with respect to one another, and if the spacecraft was at these particular points at these particular times, then that would explain why the reference points appear in the positions that they do in these pic-

*In 1999, NASA's *Mars Climate Orbiter* demonstrated what happens when things don't go well. When reporting its thruster firings the spacecraft's software used metric measurements (Newton seconds). The software on Earth thought that these reports were in pound (thrust) seconds, a smaller unit, and thus underestimated the effects of the thruster firings. This meant that JPL's model of the *Climate Orbiter*'s position became increasingly inaccurate and, when its controllers tried to inset the spacecraft into orbit around Mars, it ended up deep inside the atmosphere and burned up.

tures." That optimal arrangement of reference points was the control net.

Once Davies and his colleagues provided the control net, it could be used to position all the rest of the data. It became possible to say quite accurately where things were with respect to the planet's poles and its prime meridian. Indeed, one of the primary functions of the control net was to define the planet's latitude and longitude system—which is why Davies, as both maker of the control net and a member of the International Astronomical Union committee responsible for giving names to features on other planets, was able to put Mars's Greenwich in a little round crater within the larger crater that was being named after Airy.

After his first work on Mars, Davies did his bit in the mapping of more or less every solid body any American spacecraft has visited. By the 1970s he had completely forsaken the black world of spy satellites for the scientific delights of other planets and the personal pleasure of exploring this one: Once unencumbered by security clearances and the knowledge they bring, he was free to travel to meetings all around the world, and did so with Louise and alacrity. He's never made headlines—I doubt he'd want to—but his contributions have been vital prerequisites for much of the work that has.

But there's still more that Mert would like to do. The mathematics of the control net maximizes its self-consistency, not its accuracy. This makes it likely that it contains errors. If you had some independent way of checking it—if you had a point in the control net the location of which you knew independently—you might be able to do something about that.

In principle, such independent measurements are possible. When I interviewed Davies in his office at RAND in December 1999, America had landed three spacecraft on the surface of Mars—the two *Viking* landers in 1976, and *Pathfinder* in 1997. The radio signals sent back from those spacecraft revealed their positions very accurately with respect to the fixed-star reference system used by astronomers. If you could find the spacecraft in images of the Martian surface that also contained features tied into the control net, you could check the position of the spacecraft with respect to the net against its absolute position as revealed by the radio signals.

That would allow you to calibrate the net with new precision. Do the same for a few spacecraft and you could tie the thing down to within a few hundred yards, as opposed to a few miles.

The frustration is that you can't see the spacecraft. About the size of small cars, from orbital distances—hundreds of miles—they are lost in the Martian deserts. The Mars Observer Camera (MOC), part of the *Mars Global Surveyor* spacecraft, has been trying to pick out some sign of the three spacecraft since 1997. It is by far the most acute camera ever sent to Mars. But even MOC can't pick out the landers. Mankind has made its mark on Mars—but that mark has yet to be seen.

Lacking any proper sightings, checks on the control net using the landers' locations have had to be indirect. From matching the features that the landers see on the horizon around them with features visible in pictures taken from orbit, it's possible to make estimates of where the landers are, estimates that are potentially very accurate. Unfortunately, the different experts who try this sort of triangulation get different answers. When Mert and I met in 1999, various inconsistencies had convinced him that one bit of data that he had thought pretty good, and that he had used to calibrate the control net—a two-decade-old estimate of where exactly in the rubble-strewn plains of Chryse *Viking 1* had landed—was, in fact, wrong. In a week's time he was going to go and tell the American Geophysical Union's fall meeting about the mistake and the fact that it had introduced an error of a fraction of a degree into the control net's definition of the prime meridian. But if that was an irritation, there was also a new hope. The very next day, a new lander would be setting itself down on the Martian surface, giving MOC another man-made landmark to try to pick out. A steeple to navigate by.

The *Polar Lander*

I can think of nothing left undone to deserve success.

—Robert Falcon Scott, diary entry,
November 1, 1911

On the morning of that next day, Friday, 3 December 1999, JPL in Pasadena is awash with visitors, just as it always is when one of its spacecraft is about to do something exciting. The road leading past the local high school and up to the lab is lined with outside-broadcast vans. Inside, the tree-lined plaza at the lab's center—the place where, at the celebration to mark *Voyager 2*'s successful passage past Neptune, Carl Sagan danced with Chuck Berry—is filled with temporary trailers in which the working press will work, when there is work for them to do. It's not just journalists who are wandering around looking for gossip, coffee, and companions unseen since the last such event. There are VIPs from the upper echelons of NASA and beyond, distinguished visitors from other research centers, the families and friends of people involved in the mission. And back down the freeway at the convention center in downtown Pasadena there are hundreds of paying customers turning up for a parallel popular event held by a group called the Planetary Society, a planetary-science fan club and lobbying organization created by Bruce Murray, Carl Sagan, and a one-time JPL mission planner named Lou Friedman. The Planetfest gives the public a chance to watch the events on Mars played out on vast TV screens, to hear the findings analyzed by experts, to meet their favorite science fiction authors, to admire and buy art inspired by planetary exploration, to collect toys and gaudy knickknacks and to party the weekend away.

No other scientific event—not even the sequencing of a particularly juicy microbe or chromosome—gets attention like this. But then no other science stirs the emotions like planetary science.

The absent star of the show is the *Mars Polar Lander.* A life-sized stand-in sits in a sandbox in the middle of the plaza at JPL, a backdrop for TV reporters from around the world. Like most spacecraft, it looks a little ungainly: three widely spaced round feet, each of them braced by a set of three legs; segmented solar panels to either side, partly folded out flat, partly flush to the spacecraft's sloping shoulders, tilted to catch the beams of a sun low on the Martian horizon; spherical propellant tanks and rocket nozzles sit in its belly; antennae, masts, and a sort of binocular periscope perch on its back. A scoop on the end of a robot arm scratches the pseudo-Martian sand.

The real *Polar Lander,* cameras and legs and solar panels tucked into an aeroshell that will protect them from the atmosphere, is falling toward Mars at about 14,000 miles an hour. The last course corrections were made early in the morning, fine-tuning the trajectory to maximize the chances of hitting the chosen landing site about 600 miles from the south pole of Mars. They seem to have worked; the trajectory appears as true as if the spacecraft were running on tracks. Anyway, nothing more can be done—as Apollo astronaut Bill Anders remarked when the third stage of his *Saturn V* put him and his crewmates on course for the moon, "Mr. Newton is doing the driving now." The spacecraft has nothing to do but obey the law of gravity. Oh, and to fire the occasional rocket, discard its heat shield at the appropriate time, deploy a parachute or two, all things that have to happen precisely at the right time and can't be controlled from Earth because it would take the commands fourteen minutes to get to Mars. Standard spacecraft stuff—only nothing on interplanetary spacecraft is standard. You can never be sure you've checked out all the systems and you never fly exactly the same model twice. Every mission is a sequence of hundreds of events controlled by thousands of mechanisms and circuits, any one of which could go wrong.

Because of all this—and especially because the lab's previous Mars mission, *Mars Climate Orbiter,* ended in ignominious failure just a few months ago—the tension back at JPL is tangible. But it is also unfocused. There is no more to see than there is to do. An oddity of space exploration is that only very rarely do you get to see the process in action. You see the results, which are often spectacular in and of themselves, but there's never a cutaway camera angle to let you see the spacecraft through which these wonders of the universe are being revealed. And while it's hardly surprising that we can't see the means by which—through which—we're witnessing these wonders, it's also a great pity. You don't have to be Mert Davies, intent on refining his control net, to want to see a picture of a spacecraft on the rubble-strewn plains of Mars. You just have to be human and to want to see something human in that great emptiness where nothing human has been seen before. Such a sight would close some sort of cognitive circuit; it would make Mars a distant mirror in which we could see something of ourselves reflected. It would thicken the connections between our planets and draw Mars further into our world.

This need to close the loop explains why the most popular unmanned space mission ever was the 1997 *Mars Pathfinder.* Anyone with a Web browser could watch as its limited little rover, *Sojourner,* fitfully explored the rock garden it had been landed in. It explains why the artists displaying their wares to the faithful down at Planetfest in Pasadena do not, for the most part, just paint spectacular landscapes when they paint Mars—they paint landscapes with human participation inside them: an astronaut, a rover, even an unmanned craft. One of the most popular pictures of Mars ever painted is *Return to Utopia* by Pat Rawlings, which shows a future astronaut planting a flag—whose? We can't see—next to the second *Viking* lander, simultaneously celebrating its far-flung location and pulling it back from nature into the human world.

Here's what we're not seeing by around lunchtime on 3 December: About ten minutes before it hits the atmosphere, *Mars Polar Lander* begins making its final preparations, resetting its guidance systems, prepping one of its cameras. Mars is vast in its sky, only a few thousand miles away, half in shadow, half in sunlight, its surface a range of browns and yellows, the red of its earthly appearance revealed from space as an atmospheric illusion. At this range you can see the craters, the streaks of dust blown by the winds, the strange changing textures of the surface, the largest of the ancient, dried-out valleys, perhaps the wispy whiteness of high dry-ice clouds. New features stream around the curve of the planet as the spacecraft catches up with its target, its trajectory taking it south and east at four miles a second toward the harsh brightness of the southern polar cap. Six minutes before atmospheric entry, the spacecraft twists around so that its aeroshell heat shield is pointed forward. A minute later a set of six explosive bolts is detonated and the lander slips away from the cruise stage that has been providing it with power and communications on the eleven-month journey from Earth. From now on all the power comes from the batteries and no communication is possible until the lander's own antennae are deployed on the ground. Once the cruise stage and the lander are safely separated, the cruise stage goes on to release two microprobes called *Scott* and *Amundsen*, spacecraft designed to survive smashing into the planet's crust at high speed and then measure the moisture of its soil. They are so tiny that you could cup one in your hands like a grapefruit.

As lander, cruise stage, and probes drift away from each other, perspectives alter. Mars stops being a vast wall in front of the spacecraft and becomes a strange new land below them; the ice-white limb of the planet barring the sky becomes a curved horizon. The outer reaches of the atmosphere begin to stroke the lander's protective aeroshell, too thin at first to have much effect, but getting thicker by the second. Soon, onboard accelerometers decide the breaking force is getting strong enough to be worth bothering about and tiny thrusters start firing to keep the aeroshell's blunt nose cone pointed the right way. The atmosphere's grip tightens further. Within a minute or so, the deceleration is up to 12g—the sort of force you'd feel if a cruising airliner came to a full halt in a couple of seconds. The nose cone is at 3,000°F and the air around it is incandescent. The *Polar Lander* is a minute-long meteor in the Martian sky.

Three minutes after atmospheric entry begins, the worst is over, though the lander is still moving at 9,300 miles an hour. A gun at the back of the aeroshell fires out a parachute and the thin air rips it open at a little over four miles above the surface. Ten seconds later the charred front of the aeroshell is jettisoned and a camera pointing downward starts to take pictures of the landscape below as it rushes downward. If they make it back to Earth, these descent images will make quite a movie.

While all this is happening I'm picking at a tuna sandwich in the JPL cafeteria. I chat with some of the scientists from other projects who are gathering round the television monitors that show what's happening in mission control, then wander back across the plaza, past the model in the sandbox, to the press room. There's no hurry—the probe is silent during the landing sequence and is only due to pipe up twenty-three minutes after touchdown. Even then there will be a fourteen-minute delay as the radio waves creep across the solar system at the speed of light. Plenty of time.

Over one hundred and fifty million miles away, *Mars Polar Lander*'s legs snap out from their stowed position, ready for the ground below.

Four months later a board of inquiry decided that this was the crucial moment. When the legs snapped into position, they apparently did so with a touch more vigor than was necessary, flexing a little against the restraints meant to hold them in position. Little magnetic sensors in the spacecraft's body seem almost certain to have interpreted this flexing as meaning that the legs had encountered resistance and were bending under the weight of the spacecraft—just as they would at the moment of touchdown. The state of these sensors was being monitored a hundred times a second by the part of the spacecraft's software that was in charge of turning off the engines straight after landing and, since the legs took more than a hundredth of a second to reach their proper position, the sensors reported that the spacecraft seemed to have touched down on two successive checks. If it had heard this report only once, the software in charge of turning off the engines would have ignored the reading as a transient glitch. Hearing it twice, the software in charge of turning off the engines after touchdown concluded that the spacecraft had indeed touched down. Unfortunately, it was still two and a half miles up in the air.

A bit more than a minute later, when the spacecraft's radar said that it was only forty yards above the surface, the misinformed software had its virtual hand put on the virtual switch that controlled the engines. It turned them off straight away, unable to know or care that the spacecraft was still moving at almost thirty-one miles an hour. After falling that last forty yards, *Mars Polar Lander* hit the surface at something like fifty miles an hour, a speed it could never survive.

Back in December, no one knows any of this. About an hour after lunch on Friday, we know that the first transmission from the surface hasn't happened, but though that's a little disappointing, no one is really worried. Spacecraft are programmed to be flighty things and at the slightest sign of something out of the ordinary they are apt to go into "safe modes," which means shutting down all nonvital systems for a set amount of time. The lander's ability to "go safe" had been turned off during the descent sequence—when willful inactivity would have been fatal—but once it got down to the ground this override would turn itself off and the spacecraft would be free to go into a silent funk if some subsystem or other had exceeded its safety levels during the landing.

Over the next few days the silence gets worse. *Scott* and *Amundsen*, the ground-penetrating microprobes, are never heard from at all. To this day no one knows what happened to them. The team running the lander itself methodically lists the things that could be stopping the probe from communicating and tries to work its way around them, using various different types of radio command. Is the main antenna facing the wrong way? Then send the lander instructions to scan its beam across the sky. Did it not hear those instructions? Send them over another frequency. Did it go into a different sort of safe mode, or go safe twice? Listen at the later times when it was meant to transmit. Each possibility is a branch on what the engineers call a fault tree, and every branch has to be checked out.

While all this is going on up at JPL, down at the Pasadena convention center the Planetfest rolls on. The fact that there are no neat new pictures of the surface to be seen puts a damper on it, to be sure—but not too terrible a one. People still come to hear the assembled luminaries talk about the great future of Mars exploration. They hear from astronauts and scientists and engineers and *Star Trek* actors and Bill Nye the Science Guy, proselytizer by appointment to PBS. And they hear from the science fiction writers. From Larry Niven, who has just written a fantasy in which all humanity's dreams about Mars come true at the same time; from Greg Bear, whose *Moving Mars* imagined the planet's future as a backwater from which settlers watch the ever more high-tech Earth redefine what is human; from Gregory Benford, whose *The Martian Race*, published this very weekend, sets a new standard of technical accuracy for first-mission-to-Mars stories. And from Kim Stanley Robinson, whose books *Red Mars*, *Green Mars*, and *Blue Mars* provide the fullest picture yet attempted of life on that planet. Unlike every previous generation of science fiction writers, these men have had data from Mars's orbit and the Martian surface on which to base their visions, and they are scrupulous in their use. In their hands, the physical facts of planetary science and the romance of travel to other worlds are brought as close as they yet can be.

Meanwhile, up at JPL, what seemed so close is slipping away. After each new attempt to make contact an ever more despondent flight team comes out to face an ever smaller press corps and tell us that nothing was heard. They were so excited on Friday morning—by the early hours of Sunday, some are almost in tears. On Monday morning most have had a chance to rest, but though the faces are fresher and the eyes clearer, a certain resignation has settled in. By Monday night, all the one-fault branches on the fault tree have been evaluated; it's clear that at least two separate systems must have failed. The team will keep climbing ever more unlikely limbs of the fault tree for a week or so yet, but for the rest of us that's it. The lander is lost. The last tents in the media caravan are folded up just after midnight; we don't even have the ingenuity, or stamina, to find a bar.

Mariner 9

I think it's part of the nature of man to start with romance and build to a reality.

—Ray Bradbury, *Mars and the Mind of Man*

Mars Polar Lander was JPL's thirteenth mission to Mars and its fifth failure. *Mariner 3* died with its solar panels pinned to its side by the wrapping in which it had been launched in 1964; *Mariner 8* fell into the Atlantic in 1971; *Mars Observer* exploded as it was trying to go into orbit around Mars in 1993; *Mars Climate Orbiter* burned up in the atmosphere in 1999; *Mars Polar Lander* made its mistake just forty yards up a few months later. An optimist might point out that each got closer to the target than the previous failure. A pessimist might point out that the frequency of failure seems to be on the increase.

It's hardly surprising that, with so few missions, everything that has not been a failure has been counted a terrific success. Mars exploration is still too new for there to have been any hey-ho, business-as-usual missions. But among all these successes one stands out: *Mariner 9. Mariner 9* was the first American spacecraft to go into orbit around another planet. It was the first interplanetary probe to send back data in a flood, rather than a trickle. It was the first mission to Mars to provide images of the entire surface and record the full diversity of the landscapes. It was the first spacecraft to see a planet change dramatically beneath its eyes, to watch weather on another world. *Mariner 9* revealed a Mars that was fascinating in its own right, rather than disappointing in the light of previous earthly expectations. And *Mariner 9* allowed a small team of artists and artisans to make the first detailed, reliable maps of another planet.

There were two big differences between *Mariner 9* and its earlier siblings (two of which, *Mariner 2* and *Mariner 5*, went to Venus, not Mars). One was that *Mariner 9* had a largish rocket system on board, its cluster of spherical fuel tanks hiding the distinctive octagonal magnesium body that all the *Mariner* family shared. This engine was needed to slow the spacecraft down when it got to Mars, thus allowing it to go into orbit round its target rather than flying past it at breakneck speed, as the previous probes had. The other, less visible, difference was that *Mariner 9* would have the opportunity to send back serious amounts of data.

When *Mariner 4* flew past Mars in 1965, it seemed extraordinary that the signal it sent back could be heard at all. *Mariner 4*'s radio transmitter had a power of ten watts; it had to send data back to a target—the Earth—much less than an arc minute across (an arc minute is a sixtieth of a degree). Only a small fraction of the spacecraft's ten-watt beam actually hit the Earth, and only one ten-billionth of that fraction hit the actual receiver—a steerable radio telescope 223 feet in diameter built specifically for the Mars missions at a site a couple of hours' drive into the Mojave Desert from JPL. But the power of electronic engineers to decode such staggeringly faint signals has been one of the least celebrated wonders of the space age.* It's an ability at least as wonderful as that of actually launching things into space, and compared with rocketry it's both grown in capability far faster and been a good sight more dependable. That seventy-meter Goldstone dish in the Mojave, along with companions near Madrid and Canberra, now brings data back from the edges of the solar system, a hundred times farther away than Mars, and handles data rates as high as 110,000 bits per second. Even in the early days of *Mariner 4* the limiting constraint on the rate at which data could be sent back was not the radio link, but the speed at which the tape recorder that stored the data on board the spacecraft could play it back. And that was staggeringly slow: eight bits per second. It took weeks to send back data recorded in minutes.

*The excellent Australian film *The Dish* goes some way to redressing this oversight.

Mariner 4's pictures each contained less than a thousandth of the data in a nine-inch aerial photograph. The frames were just two hundred pixels wide by two hundred pixels deep; the brightness of each pixel was recorded as six bits of data, providing sixty-four gradations of tone between black and white. The total amount of data in every frame (thirty kilobytes) was just a little bit more than the amount of disk-space taken up by an utterly empty document in the version of Word with which I am writing this book. In principle I could download the equivalent of *Mariner 4*'s entire twenty-two-image data set from the Internet in a matter of seconds using my utterly unexceptional modem. In 1964, though, it took eight hours to get each picture back to JPL. The process was so slow that the waiting scientists printed out the numerical value for each pixel on a long ribbon of ticker tape, cut the ribbon into two-hundred-number-long strips and then colored each pixel in with chalk according to its numerical value. Every two and a half minutes another strip could be added to the picture. The first space-age image of Mars, taken by the first entirely digital camera ever built and transmitted over a hundred million miles of empty space, was put together like an infant school painting-by-numbers project.

By the time *Mariner 6* and *Mariner 7* flew past Mars in 1969, communications were far faster (though the onboard tape recorders, which outweighed the cameras whose data they stored, were still a problem). Each of the 1969 *Mariner*s returned a hundred times more data to Earth than *Mariner 4* had four years earlier. In 1971 *Mariner 9*—with a data rate two thousand times that of *Mariner 4* and a year in which to transmit, rather than a week—did a hundred times better still. And this meant that the whole scale of the operation was different. The "television teams"—so called because their instrument was basically a TV camera—on *Mariner*s *4*, *6*, and *7* had been small: Leighton, who masterminded the camera design; a few other Caltech faculty members; some JPL people; and a few select outsiders, such as Mert Davies. But *Mariner 9* was going to provide far more data than such a team could digest, and the data were to be used not just for analytical science but for the practical business of mapping. Among other things, America was committed to landing robot probes on Mars to look for life in 1976. Those probes—the

Vikings—needed landing sites, and choosing landing sites required maps.

NASA would have been happy to make the maps itself. But in the mid-1960s Congress noticed that almost every government agency had its own mapmakers and decided that the money-hungry, fast-growing space agency would be an exception to this rule. So the mapping of the planets was instead made the duty of the United States Geological Survey (USGS). This was not entirely arbitrary; the USGS already had an astrogeology branch, headquartered in Flagstaff, Arizona, that was deeply involved in the study of the moon and was helping to train the Apollo astronauts. The USGS gave primary responsibility for its study of Mars to a team of five geologists, three from Flagstaff, two from the survey's California center in Menlo Park, south of San Francisco. The senior member of the USGS team was a man called Hal Masursky; in part because Murray was at the same time working on a mission to Venus and Mercury, Masursky became one of the television team's two principal investigators (PI). The other PI was a young man called Brad Smith, a highly rated expert on Mars as observed through telescopes, who had yet to complete his doctorate.

Up to the point when he had joined the astrogeology branch in the early 1960s, Hal Masursky's career had not been stellar. He had never completed his Ph.D.; his terrestrial work had been uneventful. But Masursky became fascinated by the possibilities of geology on other worlds, and turned out to be a great success at it. The success lay not in his own scientific work—though he was a perceptive observer, his complete inability actually to write things up was something of a limitation—but in his ability to get things done within the sometimes bureaucratic world of space exploration and to explain these achievements to the world at large. Some of his colleagues considered him as vivid an off-the-cuff communicator as Carl Sagan.

Hal was at the same time a bright spark and a consummate committeeman. He was charming but dogged, willing to get down into the details of sequencing spacecraft maneuvers and download times whenever necessary, but also keeping a clear eye on the overall objectives. His astrogeological life became in large part devoted to the teamwork necessary for planning and running space missions, and he played a role in almost every major mission of the 1970s and 1980s, making sure they would send back pictures geologists could make use of. If Hal was on a committee, a planetary scientist who learned the political ropes back then once told me, the committee would get things done; if he wasn't on a committee, then you didn't want to be on it either. It was probably not an important one, and it might well not get anywhere.

Masursky was good at getting committees to work; in his personal life his gift for structure was less evident. Committee work meant he was endlessly traveling. (It's said that at times he lived in Flagstaff without a car of his own, preferring simply to rent one when he flew in just as he would anywhere else.) His ability to keep projects he was administering within budgets was famously poor. He was married at least four times, religious, and passionate in argument. He was diabetic, but rather than accepting the discipline of managing the condition he let his team do so for him. Jurrie van der Woude, an image-processing specialist then at Caltech and later at JPL, remembers finding Masursky passed out on the floor of his office late one night during the *Mariner 9* mission. Jurrie shouted for help and people came running—people already armed with candies and orange juice, because they knew what to expect. "From that point on I was part of the club. No matter where you went around the lab you'd carry orange juice with you. Nobody talked about it, but in press briefings there'd be four or five of us like secret servicemen, waiting and watching for the right time to bring him orange juice. He had this kind of a smile and every so often you'd realize that behind it he was just gone." Eventually diabetes took its toll; in the late 1980s Masursky sickened, dying in 1990. During his sad decline, he would occasionally elude his last, devoted wife and wander off to Flagstaff's little airport, sure he should be going somewhere. Now he has a crater on Mars: 12.0° N, 32.5° W, a hole seventy miles across in the region called Xanthe Terra.

When *Mariner 9* set forth from Earth in 1971, no one had seen Xanthe in close-up. No one had seen the crater that would one day be named for the principal investigator on the television team, or the striking channel that runs next to it and quite probably once filled it with water, Tiu Vallis. No one knew that Mars offered such sights. *Mariner 4* had seen a moonlike surface covered in craters. It had measured the atmospheric pressure as being much lower than most measurements from Earth had suggested—about 1 percent of the pressure at sea level on Earth. The long-held picture of Mars as a basically earthlike if very marginal environment—something like a cold high-altitude desert, except worse—was demolished. The surface had to be very old to have accumulated so many craters; the atmosphere must always have been very thin and free of moisture for the craters not to have eroded away. From the composition of the atmosphere—95 percent carbon dioxide—and measurements of its temperature and pressure—both low—Leighton and Murray had been able to predict that the polar caps, which earthbound observers had seen as water ice that might moisten their imagined earthlike desert, were in fact made of frozen carbon dioxide. *Mariner 7* seemed to confirm this theory when it passed over the south pole carrying infrared instruments capable of measuring the surface's temperature and composition.

Admittedly, Mars was not all craters. *Mariner 6* had seen that Hellas, known as a large bright region to the earthbound astronomers, was much smoother than the cratered terrain next to it, though no one could say why. The same spacecraft also sent back pictures of an odd terrain quickly termed "chaotic," a collapsed jumble of a landscape from which a few tabletop mesas stood proud. It was as though the land had rotted from within. But though such features might prove interesting, the general impression was of a dull, geologically inactive place, more or less unchanged since the creation of the solar system, a place little more interesting than the Earth's moon and far harder to get to. Bruce Murray, who unlike many in the business had never had a boyhood romance with the stars, took a certain delight in debunking the delusions of people who still wanted to think of Mars as at least a little earthlike. Murray has a certain intellectual aggression, as do many Caltechers—the USGS geologists on *Mariner 9* used to be amazed by the frequency and ferocity of the arguments that Murray's students on the team, Larry Soderblom and Jim Cutts, would get into. Nostalgic notions of an earthlike Mars gave Murray's belligerence its *casus belli*. Mars was simply not what people had thought it to be. Rather than a world to be experienced in the imagination, it was a planet to be measured, a planet in the new space-age meaning of the term, something woven from digital data streams and ruled by the hard science of physics and chemistry.

On 12 November 1971, the night before *Mariner 9* was to go into orbit, Caltech held a public symposium on "Mars and the Mind of Man" featuring Murray, Carl Sagan, and the science fiction authors Arthur C. Clarke and Ray Bradbury: It was the genteel ancestor of the bigger, brasher Planetfests that accompany today's missions. Murray cast himself in wrestling terms as "the heavy—the guy with the black trunks." He acknowledged people's "deep-seated desire to find another place where we can make another start . . . that is not just a popular thing [but] affects science deeply." He then set about using his experience of *Mariners 4, 6,* and *7* to pour cold water—in fact frozen carbon dioxide—on such fancies. Carl Sagan, a new member of the television team and already a passionate advocate of the search for life in planetary exploration, responded by saying that nothing seen so far had ruled out life on Mars—it had just made it harder to imagine if you were parochial enough to imagine all life must be like Earth life. Clarke optimistically suggested that if there weren't life on Mars in 1971, there certainly would be by the end of the century.

While Clarke and his colleagues spoke in Caltech's auditorium, events up at JPL were turning out quite dramatic enough without any added fiction. One of the reasons that 1971 was a good time to launch the first Mars orbiters was that Mars, which has a markedly eccentric orbit, would be at its closest to the sun at the time when it was most easily reached from the Earth. Unfortunately, perihelion warms the Martian atmosphere up quite a lot and the resultant winds can kick up dust storms. This possibility had been discussed earlier in the year by the *Mariner* mission operations team. Brad Smith, Masursky's partner at the helm of the television team, said it would not be a problem. But Smith was wrong. The great storm started on 22 September 1971. Within a few days almost half the southern hemisphere was obscured by the brilliant cloud and a week later a second storm started farther to the north. Soon the storms merged. Telescopes on Earth saw a Mars utterly without features—and so did *Mariner 9.* Its first pictures, sent back on 8 November, revealed no detail whatsoever—wags joked that they had arrived at cloud-covered Venus by mistake. On 10 November, when the pre-orbital images should have been as good as those from *Mariners 6* and *7,* all that could be seen was the faint outline of the south polar cap and a faint dark spot. It turned out to correspond to the location that Schiaparelli had called "Nix Olympica"—the Snows of Olympus. Two days later three more dark spots were seen a thousand miles or so from Nix Olympica, forming a line from southwest to northeast across the region called Tharsis. The rest of the planet was still completely blank.

Two days later, after the spacecraft had gone into orbit, new pictures revealed that each of these spots had a crater at its center. Carl Sagan took a Polaroid of the computer screen and rushed to the geologists' room. Masursky and his colleagues immediately realized what they were seeing. These were not impact craters like those seen by the previous *Mariners*, but volcanic calderas. Nix Olympica and the other features—dubbed North Spot, Middle Spot, and South Spot—were volcanoes, volcanoes vast enough to stick out of the lower atmosphere into air too thin to carry the fine Martian dust. Within hours, Masursky was telling the waiting press corps all about it. Murray, who as well as sporting the black trunks of the killjoy was taking on a role as the television team's prudent conscience, was aghast. Mars had previously shown no signs of volcanism; it was surely rash to jump to such a dramatic conclusion. But within days more detailed photos showed without doubt that Masursky was right.

It's easy now to scoff at Murray's reluctance to see the truth. Mars's volcanoes have become, along with its vast canyon system, the things for which the planet is best known. Inasmuch as there is a popular picture of Mars today, these features—four big lumps with a long set of deep gashes to one side, rendered in a reasonably garish red—are what make it up. In some ways, though, Murray's reluctance to credit such things seems almost fitting, a greater tribute to their stature than straightforward acceptance. It may sound like a lack of imagination—but if you wanted to, you could read it as the opposite. Maybe Murray had the imagination to look beyond the simple images of calderas and see how quite dauntingly huge the volcanoes would have to be in order to show up on *Mariner 9*'s pictures of a planet wrapped in dust from pole to pole.

Think of the commute that some of the USGS astrogeologists were making on a weekly basis between San Francisco and Los Angeles; like a few thousand people every day, I made it myself while researching this book. You come off the tarmac at San Francisco airport and wheel around over the South Bay; northern California drops away beneath you, views open up. By the time the plane is at its cruising altitude of 33,000 feet, the view has spread out across the state. The Coast Range beneath you is a set of soft creases in the Earth's crust, the Sierra Nevada a white rim on the horizon. After about half an hour's flight at a fair fraction of the speed of sound, you start to drop down and pull out over the Pacific, then come back around into LAX. And if your plane could fly through solid basalt, that entire flight profile would fit easily inside the bulk of the volcano then known as Nix Olympica and now called Olympus Mons.

Olympus Mons is a softly sloping cone sitting on a cylindrical pedestal, a flattened lampshade on a 70mm film canister. The face of the pedestal is a cliff that circles the whole mountain and rises on average three miles or so above the surrounding plain. Stick that pedestal onto California and it would cover the center of the state from Marin County in the north to Orange County in the south. The mountain's peak, more than ten miles above the top of its surrounding cliff, would be high in the stratosphere, far above the reach of any passenger jet. You would be able to see it halfway to Flagstaff, a gently humped impossibility peering over the western horizon.

Yes, Olympus Mons is a mountain, built up by eruption after eruption of smooth-flowing basaltic lava. Yes, Earth's "shield volcanoes"—like Ararat in Turkey, or Kilimanjaro in Kenya, or Mauna Kea in Hawaii—were built in a similar way and have much the same profile. But the scale of the thing is incomparably grander. Mauna Kea, Earth's biggest volcano, would fit into the huge crater at the summit of Olympus Mons with room to spare. If you strung the arc of Japan's home islands round its base the two ends wouldn't meet;

nor would the peak of Fuji clear the top of the great cliff that they were failing to encompass. An Everest on top of Everest would not come to the summit of Olympus Mons.

This single brutish Martian lump is larger than whole earthly mountain ranges. Its bulk—some three and a half million cubic kilometers of rock—is about four times the volume of all the Alps put together. If you wanted to build one on Earth, you'd have to excavate all of Texas to a depth of five miles for the raw material—and you'd still be doomed to failure, because the planet's very crust would buckle under the strain.

North Spot, Middle Spot, and South Spot, stretched out along the ridge of Tharsis, are smaller than Olympus Mons. But not by much.

The great storm, rather than obscuring Mars completely, had in fact served to highlight its most dramatic features. It also set Sagan—always alert for lessons from other planets with relevance to this one—to wondering whether similar phenomena might have any relevance to the Earth. *Mariner 9*'s infrared spectrometers showed that the dust did not just obscure the Martian surface from earthly eyes; it also chilled it by shielding it from the sun. In 1976 Sagan, his student James Pollack, and other colleagues produced papers showing how the dust thrown into the Earth's stratosphere by large volcanic eruptions could cool the home planet in a similar way. Such cooling was to be put forward in the early 1980s as the mechanism by which a large impact by an asteroid or comet—an event guaranteed to kick up a lot of dust—might have killed off the dinosaurs. This new mechanism for mass extinction led to Pollack and his colleagues being asked to model the sun-obscuring effects of nuclear war, and thus to the idea of "nuclear winter." Having gone to Mars to look for signs of life, Sagan found intimations of planetary mortality.

As the cooling planet-wide pall of dust started to ebb down the volcanoes' flanks in late 1971, the television team began to pick out the outlines of other features: Depressions, in which there was more airborne dust to reflect sunlight back into space, started to stand out as bright blotches. By the middle of December a vast bright streak had become visible to the east of the three Tharsis volcanoes. When the dust had settled out further, the streak was revealed to be a set of linked canyons more than a thousand miles long and three miles deep. It would come to be called Valles Marineris after the spacecraft through which it was discovered. By the time the dust subsided in 1972, large parts of the planet's northern hemisphere had been revealed as plains much more sparsely cratered than those over which the first three *Mariner*s had passed. At the same time, other features known from earthly observation, like bright Argyre and Hellas, turned out to be the remnants of absolutely vast impacts.

Most striking of all, particularly to Masursky, were the erosion features. In some places long, narrow valleys ran for hundreds of miles across the plains with few, if any, tributaries. In other regions there were branching networks of smaller valleys, suggestively similar to those that drain earthly landscapes. And elsewhere there were vast, sweeping channels that seemed to have torn across the crust with unbelievable force, scouring clean areas the size of whole countries. Had water done this? Masursky seemed sure of it and waxed lyrical on the planet's lost rains to journalists; Murray looked on, grinding his teeth. After all, this was an alien world of new possibilities. Streams of lava might have been responsible—or torrents of liquid carbon dioxide, or gushing hydrocarbons, or slow-grinding ice. Even the thin winds were suggested as possible scouring agents—and though that was a spectacular stretch, it was increasingly clear that wind did indeed play a large role in the way the planet looked. Everywhere there were streaks where dust had revealed or hidden the surface beneath; in some places there were full-blown dune fields. The seasonal changes observed from the Earth and held by some to mark the spread of primitive vegetation—changes that would have been *Mariner 9*'s primary focus, had its sister ship, *Mariner 8*, not fallen into the Atlantic just after launch and thus bequeathed the main mapping mission to its sibling—were now explained by the wind, at least in principle.

And there was yet more for Masursky and Murray and their colleagues to wonder at and argue over. Strange parallel ridges and lineations running in step for hundreds of miles, the collapsed chaos features seen by *Mariner 6,* which now appeared to be sources for some of the great channels. Rippling bright clouds of solid carbon dioxide (such clouds, streaming off the heights of Olympus Mons, provided the intermittent bright white expanses that made Schiaparelli think of snow and call the area Nix Olympica). Most strikingly, there were regions at the poles where the interaction of wind-borne dust and expanding and contracting polar caps had built up a weird, laminated terrain. Each layer must correspond to a different set of conditions—different wind patterns, different climates. Millions, maybe billions of years of history were there in those layers, just waiting to be read if only you could get to them and figure out what made them. Murray, in particular, found these polar layered terrains fascinating. Thirty years later he still does. He was to be part of the science team on the ill-fated *Scott* and *Amundsen* microprobes that accompanied *Mars Polar Lander.*

The twenty-four people working shifts on the television team had more than enough data to keep them happy. Every twelve hours a new swathe of pictures would come back, covering a band of the planet every seventeen days. There were always new things to see, new things to think about, new things to ask for close-ups of at the next opportunity. And in the end Mars's rocky surface was stored in their computers and tacked up on their walls, almost seven gigabytes of data: 7,329 images. Mars was now much more than one of Tennyson's points of peaceful light—it was taking on, in Auden's words, "the certainty that constitutes a thing." It could be measured in detail, and properly mapped.

The Art of Drawing

How wonderful a good map is, in which one views the world as from another world thanks to the art of drawing.

—Samuel van Hoogstraten, *Inleyding tot de Hooge Schoole der Schilderkonst*
(translated in Svetlana Alpers, *The Art of Describing*)

In 1959 Patricia Bridges, a gifted illustrator with a degree in fine arts, started making maps of the moon for the Air Force Chart and Information Center (ACIC) in St. Louis. Her technique soon established ACIC as a better moon-mapping outfit than its great rival, the Army Map Service. But St. Louis was not a particularly good place from which to see the moon and, though mapping from photographs was possible, direct observation was better. The ever-changing smearing of the atmosphere made it almost impossible for 1960s cameras to capture the moments of clarity in which the moon's features are best seen—but the well-trained human eye could seize such brief impressions, understand what was seen in them, and remember it. Through a good telescope eyes as keen as Bridges's could gauge lunar details as little as two hundred yards across, more than twice as acute as the resolution in photographs.

The mappers wanted that clarity and so they needed regular access to a good telescope. The twenty-four-inch telescope that Percival Lowell had built in Flagstaff with which to look at Mars was one of the best available, benefiting from high altitude, clean skies, and clear nights. So the Air Force moon mappers moved to the Lowell Observatory, settling in permanently in 1961. They were based in a small cabin—previously a machine shop and lumber store—just a hundred yards or so from the observatory's dome. By observing the same features lit from different angles on the waxing and waning moon, Bridges was able to get a sense of the features' forms that a single photograph could never give. Sometimes she would sit there working on her maps night after night until the seeing was just so, at which point a colleague inside the dome would call her on the telephone, and she would bundle up in her coat and run over to the telescope to capture some new detail of her subject.

In the mid-1960s, with the Apollo program a national priority, the Flagstaff operation blossomed. More than half a dozen cartographers were trained in Bridges's technique for lunar shaded-relief mapping. Shaded relief is a way of using heavier tones to suggest the shadows of hills and ridges on a map, giving the eye a sense of the third dimension. There are plenty of ways of doing the shading—with pencils, with paint, with chalks, even through a rather cumbersome system of embossing the relief onto plastic sheets and then photographing them lit from the side. But for the most part these are used to add shading to maps in which the relief is already clearly known through surveying, maps on which the topography is already defined by contours.

For the moon mappers the shadows with which they defined the landscape's features were not an evocative extra to ease interpretation or please the eye. They were the essence of the map, the ultimate expression of the surface's form. As such they needed to be rendered with minute fidelity, and the tool of choice was the airbrush, capable of capturing both the finest details—which is why people who retouched photographs relied on it in the days before Photoshop and similar software—and producing precisely graded washes, which was what commercial artists liked about it. There are other ways of producing maps of the planets: Using *Mariner 9* pictures and Mert Davies's control net, a British astronomical artist called Charles Cross did a very pretty and accurate set of maps with pencil and charcoal. These were used to make the first-ever comprehensive atlas of Mars, with text by Britain's leading popularizer of astronomy, Patrick Moore. Cross's work was fine; but compare it with the far greater precision of Bridges's moon work and you see immediately why, when the USGS started planning the production of official maps of Mars, the airbrush technique would have been the obvious one to use even if its leading proponents had not been located in the same town as the USGS astrogeology branch. Ray Batson, the USGS cartographer whom Masursky had chosen to run the mapmaking team, made recruiting Bridges, who had left the Air Force mappers in 1968 but still lived in Flagstaff, one of his first priorities.

Another recruit from the Lowell team was Jay Inge. Inge had been a keen stargazer from boyhood on, but bit off more than he could chew, mathematically, when he enrolled for physics and astronomy at the University of California, Los Angeles, in the 1960s. After the first semester he was "casting around for things to do" and ended up taking a degree in biomedical illustration. Then he heard from a friend—one of his childhood telescope buddies—about what was going on at the Lowell Observatory. The moon mappers needed his illustrating skills and they offered a way back into stargazing. So Inge joined the team at Lowell.

Inge augmented the techniques Bridges had developed in various subtle ways. One particular gift he brought was a dexterous use of the powered eraser, not to get rid of errors—"an eraser is never used to rescue a poor drawing," he wrote sternly in a manual on shaded relief mapping—but as a technique for highlighting things. This was, in a way, an adaptation to the airbrush of the "dark plate" mapmaking technique that was then sometimes used for charts of the ocean floor; dark plates double illustrators' options by allowing them to both add and subtract from what was on the page to begin with. By taking ink away from the airbrushed original with a trusty K&E Motoraser, the illustrator could clarify and accentuate fine details, especially in the more deeply shaded parts of the maps.

By the time they made a start on the *Mariner 9* images Bridges and Inge were highly accomplished, and the techniques they had developed for the moon were being taken up elsewhere. Inge had a fair amount of experience with Mars, too; while at Lowell he compiled telescope observations into a number of "albedo" maps that showed the light and dark markings familiar for centuries (albedo is an astronomical term for the brightness with which an object reflects sunlight). But the spacecraft data offered new challenges. The television images from *Mariner 9* were far better than any previous pictures of Mars, but they were very poor compared with the best images of the moon seen from the Earth. (Even those observations were not as good as the pictures taken by the high-resolution camera designed for national security work that flew on board the Lunar Orbiter missions, which in the late 1960s overtook airbrush work as the state of the art for lunar mapping.) And with Mars there was no running up to the telescope in the middle of the night to get a better look. It wasn't all the spacecraft's fault: Mars was not a terribly good photographic subject. Its surface was pretty uniformly dark, and even after the great storm of 1971 had died down the atmosphere carried a residual obscuring burden of dust, not to mention occasional clouds.

The pictures were a lot less than ideal. Their saving grace, though, was that they were stored in a digital format. And even in the 1970s, there was a lot you could do with digital data to make it look better. The distortions in shape and brightness due to the design of the TV tubes could be dealt with. So could the after-image effect caused by the fact that vestiges of the previous picture would be mixed in with the current one. (If all this makes the cameras sound bad, well, they were; but they were also the best that could be sent to Mars.) Contrast could be increased spectacularly with new image-processing algorithms that massaged the data so that small variations in brightness were exaggerated into large ones. The computers could also "rectify" images in which the camera had been pointed off at an angle, rather than straight down, putting them into a form suitable for mapping. Points from Merton Davies's control net would be identified in a set of pictures and a graph would be created that showed how those points would be arranged in a given map projection. Then the image files would be stretched and squashed until the control points in the images matched the pattern prescribed in the idealized graph. An easy way to check that the system was working correctly was to look at the shapes of craters before and after. In pictures the spacecraft had taken at an angle, perspective made the craters on the surface look elliptical; in pictures the computers had given a correct projection, they were circular.

This time-consuming process produced "photomosaics" with their proportions corrected and their features enhanced. But these mosaics still had their shortcomings. Some of the individual images that made them up would be darker than others, giving a sort of fish-scale effect to the assemblage. The images would also have been taken at different times of day and thus different pieces of the landscape would be lit from different directions—confusing to the inexpert eye and irritating to the expert one. Imperfections in the control net squashed and stretched some areas (in the case of the north polar region the small number of distinctive landmarks was particularly problematic, and would cause Inge no end of grief). And many useful images were simply excluded. Much of the Martian surface had been visited repeatedly by *Mariner 9*'s cameras, but only one image of any given feature could make it into any given photomosaic. The others had to be left out, even if they offered extra information. In short, even when rectified, the primary *Mariner 9* mosaics were ugly, confusing and less detailed than they could have been.

This was where the airbrush mappers came in: Bridges, Inge, and their junior colleagues Susan Davis, Barbara Hall, and Anthony Sanchez. They overlaid the photomosaics with Cronaflex, a Mylar film covered in a translucent gel onto which they would apply their ink. For the most part—different mappers had different styles—they would first trace the obvious features, such as rims of craters and edges of valleys, then start to work in the detail. As well as looking through their working surface at the mosaic beneath, they would also look at any other pictures they had that showed the same features. They built up a mental image of the forms they were trying to portray, their imaginations reaching into the images for detail, their discipline pulling them back from self-delusion.* They made their Mars in their minds and their airbrushes whispered it onto the Cronaflex. The concentration required was phenomenal. Ralph Aeschliman, the only airbrush artist still working at Flagstaff in 2000, likened it to being a bathroom plunger stuck to a television screen, "If you got interrupted there was this *schwooup* noise as you tore yourself away."

Making the maps was a way of working through the data, one that

*There were exceptions. Experience on the moon had led the mappers to treat odd features as craters until proved otherwise, so some Martian oddities ended up drawn as craters even when they weren't. One of these noncrater craters went on to feature as a landmark in a rather good science fiction novel, Paul McAuley's *The Secret of Life*.

did so in images rather than words. Inge talks of it as an act of interpretation, a way of precisely describing the television team's data. But these were not just descriptions; they were pictures. Indeed, to some they were art. Aeschliman was scraping a living as a landscape artist in the Pacific Northwest—he had an intriguing style that drew on Chinese influences—when a reawakened interest in astronomy led him to buy some of the USGS maps in the mid-1980s.* "I'd always hated airbrush art—it was always so slick—but in those maps it was like dancing. It's hard to describe—very disciplined but very free too, the representing of a mental landscape built up from source material that's very scattered and different." When his rent increased three times in a year, he decided it was time to head for warmer climes and clearer skies in the Southwest. When he got to Flagstaff, he came to the USGS and asked for a job.

Aeschliman was instructed in the planetary mappers' technique by Bridges—"There were times when I thought I'd just never be able to do it"—but his greatest respect was reserved for Inge. "He was very spontaneous. He worked very rapidly and his work sort of sparkles. It has a presence." Inge, now confined to a wheelchair by multiple sclerosis and myasthenia gravis, is flattered when I remind him that Aeschliman thinks of him as an artist. Though his living room walls are decorated with expressive abstracts he's painted, Inge claims to set little store by them. "I'm a dabbler; I don't think I qualify as anything better than a good motel artist." But then Inge didn't set out to be an artist; he was always set on being part of the research program itself. So while he plays down any pride that he takes in the obvious artistry of his maps, he is happy to boast about the projects they have made him part of. "Of the twenty-five mappable surfaces in the solar system—the solid planets and moons we've visited—I've worked on eighteen of them."

Of all those surfaces, Mars had the most time and ink devoted to it. In 1971 Batson and Masursky decided that they would cover the

*It's a nice coincidence that the father of astronomical art, Chesley Bonestell, was also much drawn to Chinese landscapes and delighted in being able to produce good enough examples of the genre to fool his friend Ansel Adams into accepting them as genuine.

whole planet at a scale of 1:5,000,000—about eighty miles to the inch, a scale at which the smallest features identifiable in the *Mariner 9* data would be just discernible. To make the work manageable, the surface was cut into thirty pieces, known as quadrangles. Pat Bridges mapped an astonishing eleven of them; Hall, Davis, and Sanchez between them did another twelve; Inge did seven, as well as maps and globes of the whole planet. He also oversaw the production process, imposing rigorous quality control, doing the halftone separations personally, flying to the survey's presses in Reston, Virginia, to supervise the printing and making "an obnoxious little shit" of himself. The series was not finished until 1979, eight years after *Mariner 9* arrived at its destination. But the final result is magical. These are maps to lose yourself in, like windows in a spaceship's floor. They seem at the same time transparent to the truth and dense in artistry. They combine the presence of that which is real with the power of that which is inscribed.

The 1960s and 1970s were a great time for mapping. The space age was coming home to roost: The Earth, that always-inhabited, always-experienced world, was being made over into an objectified planet just like its neighbors, a minutely measured ball of rock and water. In the 1960s Argon spy satellites, offshoots of the Corona program with cameras optimized for mapmaking, were used to produce vast mosaic maps of poorly surveyed Africa and Antarctica. Other satellites were busily tightening up a global control net far more sophisticated than the Martian one, refining humanity's knowledge of the shape of its world so that missiles would more easily be able to find their targets. The needs of the nuclear submarines from which those missiles would be launched, along with the interests of a new generation of Earth scientists, were driving new studies of the Earth's ocean floors; while detailed data on the ocean depths were highly classified, beautifully drawn maps based on those data allowed Earth scientists to see the spreading ridges and transverse faults central to new ideas about plate tectonics.

But the Earth, partly because of those submarine-hiding oceans, could never be mapped in its entirety in the way that Mars was. Nor could it be mapped with such supreme disinterest. Earthly maps are heavy with duties to property and strategy, duties that can warp and distort them. On Mars everywhere was alike; nowhere was rich, or strategic, or owned, and so a pure disinterest reigned. There was a political point in their publication—these were American products, based on American ingenuity, printed by the American government—but in the images themselves there was nothing but the data, the interpretation, and the artist's style.

Though they were in some sense less faithful to the truth of the planet than the television images they were based on, the maps were far more approachable, especially for the layperson.* They had a feeling of naturalism that the other forms the data were presented in lacked. Like most naturalism, this was highly contrived, depending on a number of strict conventions. Tricks of shading were used to make sure the users' eyes saw craters as dimples, not domes (an inside-out illusion endemic in photographs of planetary surfaces). The regional differences in the surface's albedo—the curves and blotches that are all that you can ever see of Mars through any earthly telescope—were suppressed. Mars's albedo was controlled not by the nature of its surface features but by the way the wind blew dust around and over them (the dusty bits were bright—the bits swept clear were darker), and winds were not something the mapping project was interested in. Inge developed a clever way of making separate albedo plates so that the maps could be printed with regional patterns or without, but after a few quadrangles the effort was given up. Nor was the color on the final prints—a soft, light-brownish pink—the real color of Mars. It was a color chosen by Inge just to give a feeling of Mars. And somehow it did. The maps are indeed, as Inge always insists, technical documents that happen to have been drawn up in pictures, not words. But they were something more, too. After the maps were made, the real Mars was not only a surface under the spacecraft's circling cameras. It was also something directly available to, and through, human minds and eyes and hands.

*The other great cartographic products of the *Mariner 9* mission, a set of ten-foot globes made at JPL through the painstaking hand-positioning of fragments of images on spherical surfaces, have a strange patchwork texture that makes them almost impossible for anyone but an expert to interpret.

Sadly, mapping Mars descended from being a delight to being a chore. Almost as soon as the first series of 1:5,000,000 maps was finished, it was decided to revise them using new pictures taken by the *Viking* orbiters, which had reached the planet in 1976. The original artwork was pulled out of storage and reworked on the basis of the new data. Because the control net had evolved, features had moved a bit and fudges had to be made. New detail was added, but in some cases the resulting maps looked cluttered and confusing. Inge was no longer checking the presses and the colors became less subtle. Frictions between Inge and Batson took their toll. Bridges retired in 1990; Inge left in 1994 and became embroiled in litigation with the Survey on the basis that his medical condition was unreasonably used to prevent his reemployment in 1997.

The airbrush artists were not replaced. Batson saw that new computer systems could make photomosaics ever more maplike—the Mars Digital Image Mosaic 1:2,000,000 series he oversaw the creation of is now the basic reference for almost everyone who studies the Martian surface. The topographic mapping of the planets is now almost entirely a matter of image processing. This has not banished beauty. In the late 1980s a geologist named Alfred McEwen produced some magnificent views of large reaches of the planet on the computer while at Flagstaff. An image he made of the western hemisphere—the ridge of Tharsis volcanoes close to the limb, the gash of Valles Marineris across the center, the thin trace of Echus Chasma running hundreds of miles toward the north like a gold highlight—may be more widely circulated than any other picture of the planet. It is to Mars what Harrison Schmitt's endlessly reproduced picture of East Africa, the Indian Ocean, and Antarctica, taken during the *Apollo 17* mission, is to the Earth. But though they can be beautiful and highly accurate—on such work you can improve things pixel by pixel if need be—the computer images lack the intimacy of the airbrush. By 2000 the latecomer Aeschliman was the only old airbrush hand remaining at the Survey's Flagstaff branch and he was doing his work entirely on screen. There is still an airbrush on the premises somewhere, but there is no longer any compressed nitrogen to bring it to life.

The maps themselves, scarred by revisions, sit in storage. All, that is, except one. Late in 1972, according to Jurrie van der Woude, who looked after some of the logistics of the *Mariner 9* pictures and has been doing similar things at JPL ever since, Bruce Murray pleaded for a copy of the one-sheet shaded relief map of the whole planet that Batson's team was making based on the *Mariner* data. Van der Woude called Batson in Flagstaff, who admitted that Inge and Bridges had finished the map. Plates of it were being made for reproduction. When it was released it would turn out to be big news—a page of its own in the *New York Times*, a British tabloid headline screaming "American Miracle—Map of Mars!" But it was not yet released. Indeed, there were not yet any printed copies.

Van der Woude persisted; eventually Batson agreed to send the original over to Pasadena, as long as it came back swiftly. Van der Woude gave it to Murray with dire imprecations that it must, but must, be returned in two days. Three days later van der Woude started to think that the normally friendly Murray was avoiding him.

It took a week or so for van der Woude to corner Murray and find out what had happened. Murray was an ambitious man; within a few years he would be the director of JPL. He had wanted the map to impress Harold Brown—then president of Caltech, later secretary of defense. Brown had thought the map wonderful and asked to show it to a guest, Henry Kissinger. Kissinger, too, was impressed and commandeered the map in order to offer it as a gift to Leonid Brezhnev; in part, we can be sure, because the Soviet Union's two missions to Mars in 1971 had failed, their preprogramming too rigid to allow them to sit out the dust storm in orbit before getting to work, as *Mariner 9* had done. And so the map had gone to the Kremlin.*

At least that's Jurrie van der Woude's story. Inge remembers that the map was lost, but not how. Murray says he remembers nothing of it—as does Harold Brown. Kissinger has proved elusive on the

*They were not the first planetary maps to get to the highest offices. Maps of the moon by the engineer James Nasmyth—better known for his steam hammer—so fascinated Prince Albert that he had Nasmyth present them to Queen Victoria, who was duly impressed.

matter. So I have to doubt it. But I want it to be true. I want the first modern map of that planet to have played a role, even just a small one, in the history of this one. I want it to have reached the top. And I want it to have ended up where Jurrie says he last saw it, glimpsed in the background during a televised interview with a Russian space scientist, apparently taking pride of place on his office wall. I want it to be somewhere where it gets treated as an icon.

The Laser Altimeter

> Then felt I like some watcher of the skies
> When a new planet swims into his ken;
> Or like stout Cortez when with eagle eyes
> He stared at the Pacific—and all his men
> Looked at each other with a wild surmise—
> Silent, upon a peak in Darien.
>
> —John Keats,
> "On First Looking into Chapman's Homer"

On 13 February 1969, nine days before *Mariner 6* set off for Mars and five months before Neil Armstrong was to step onto the dust of the Sea of Tranquillity, the newly inaugurated president, Richard Nixon, asked his vice-president, Spiro Agnew, to explore the options for a post-Apollo space program. Agnew became enthused. When *Apollo 11* made its historic landing that July, he talked of committing the nation to the goal of sending people to Mars. The report of Agnew's Space Task Group, offered to the president in September 1969, discussed this possibility and many others—but more or less ignored the question of how much it was going to cost. Nixon could not allow himself that privilege.

In May 1971, the month *Mariner 9* was launched, the Office of Management and Budget (OMB) informed NASA that its budget, already significantly cut back from its mid-1960s heights, would be frozen for five years. On 5 January 1972, two months after *Mariner 9* reached Mars, President Nixon authorized NASA to start work on a reusable Space Transportation System—the space shuttle. There was severe doubt—at OMB and elsewhere—as to whether this was wise; NASA's claims that it would make space travel far cheaper were highly dubious. But it was the least ambitious thing on offer that would keep people flying into space. And people in space, even if they had nowhere particular to go once they got there, was an idea that meant something to Nixon and to many of the men around him.

From 1972 onward the space shuttle was central to NASA's institutional survival. A national means became the agency's end. Almost everything else was either a distraction or, if it looked expensive, a threat. The planetary missions already approved—the *Pioneer 10* and *11* missions to Jupiter and Saturn, and the *Viking* missions to Mars—were not in too much trouble. But missions not already accepted were delayed and scaled back. The ambitious TOPS probes to the outer solar system that JPL had been planning were replaced with enhanced, enlarged versions of the now aging *Mariner* spacecraft design. In the end that did little harm—launched in 1977, the *Voyagers* were a spectacular success. But they were the last hurrah of the '60s horde. Between 1979 and 1991 JPL launched only two more planetary spacecraft.*

It was in this climate of cutbacks that the *Viking* landers lowered themselves to the surface of Mars in 1976. For years they sampled dead soil, analyzed dry winds, and photographed barren landscapes at two unprepossessing sites in the planet's northern hemisphere. In engineering terms they were a spectacular triumph. Their accompanying orbiters, meanwhile, added huge numbers of new pictures to the *Mariner* archive. And that was just as well, since the *Viking* treasury was to be the raw material for most of the next two decades of Mars research. The *Viking* missions were the most expensive effort in the history of planetary exploration and their single take-home message, according to most of the scientists involved, was that Mars was as lifeless seen from the surface as it had appeared to be from orbit. Expensive, dead, and already the subject of overflowing data archives, to NASA budget-setters Mars looked like a pretty good place not to return to.

That didn't mean that scientists stopped talk about new missions

*At a conference in Germany in 1990, a frustrated JPL engineer named Donna Shirley told a story about a recently deceased NASA engineer asked by St. Peter what he'd achieved with his life, to which the answer was, "First viewgraph, please . . . " Shirley eventually led the *Mars Pathfinder* team.

to Mars. At any given time there will always be lots of ideas for missions that someone or other dearly wants to see fly. Some are little more than water-cooler chatter. Some are studied but never approved. Some are approved but then dropped. Each one that flies leaves the ashes of a dozen other dreams in its wake. The field of planetary science is full of brilliant people in their forties who have still never managed to get an instrument they defined or built onto a spacecraft, never gaining the status of a principal investigator.

In the early 1980s one of the competing dreams was a mission called Mars Geoscience/Climatology Orbiter. Its proponents admitted that, yes, it did seem that Mars was a dead planet, both biologically and geologically. Although there were arguments about how to date features on the surface—arguments that will be discussed later, along with many other scientific issues some readers probably think I'm passing over too quickly at the moment—most of the interesting events in Martian history were thought to have happened billions of years ago. But dead could be interesting and besides, Mars had only been studied from a fairly narrow point of view. Most of the data were in the form of pictures. To geologists like Hal Masursky and his crew, these pictures were great. Geologists are interested in stories about which rocks are where and how they got there. While pictures taken from orbit were not terribly good guides to the nature of the rocks, their form and arrangement—the morphology of the surface—were well captured, and that provided a lot of grist to the geological mill.

Geology, though, is not the only way to study a planet. Geophysicists are interested in understanding physical forces and processes, something they seek to do in large measure by building mathematical models. From this point of view, pictures, while pretty, are no substitute for numbers. Geochemists are interested in the chemical elements from which planets are made up. Climatologists want to know whether they can understand the atmosphere's behavior. All these disciplines had an interest in Mars that the *Viking* data set couldn't satisfy. A modest orbiter dedicated to geophysics, geochemistry, and climatology might be able to fill in the gaps in humanity's knowledge of Mars—the mineral composition of its surface, its precise shape, the strength of any magnetic field, the structure of its atmosphere—with model-friendly numerical data.

The argument was pretty good, the prospective investigators were widely respected and the idea that the spacecraft could be a cheap modification of a design already used for satellites orbiting the Earth was a plausible and appealing selling point. Indeed, the idea was intriguing enough that it started to grow. If a small geosciences spacecraft could be sent to Mars, why not send a similar one back to the moon? Or to orbit an asteroid? Buying the same design and components in bulk would keep the prices down, after all. And so the geoscientists' Mars mission became *Mars Observer,* first in a new line of Observer spacecraft. Under pressure from geologists like Masursky—and with an eye to public relations—NASA added a small, comparatively cheap camera to the design; left to the geophysicists *Mars Observer* would have had no ability to take pictures in any usual sense of the word.

One of *Mars Observer*'s objectives was to get a detailed picture of the planet's relief. The *Mariner* and *Viking* scientists had used a wide number of different techniques to try to calculate how high features on the Martian surface were. They used triangulations based on the visual images. They used the precise instants at which radio signals from orbiters were cut off as they passed behind the planet. They used subtle differences in the amount of infrared and ultraviolet light reflected from different parts of the planet through different depths of atmosphere. They used narrow beams of radio waves bounced off the surface by Earth-based radio telescopes. All these different measurements were synthesized by Sherman Wu, in Flagstaff, to provide contours for the Survey's maps. But even Wu did not think the elevations he painstakingly arrived at were accurate to more than about two-thirds of a mile.

Mars Observer was to sort all this out with an onboard radar system developed by a team at NASA's Goddard Space Flight Center led by David Smith, a British geophysicist. Smith is a warm, affably excited man who, had he stayed in his native country, would be endlessly returning the smiles of women struck by his resemblance to the widely adored sportscaster Des Lynam. He had spent the 1970s applying the geophysical ideas attendant upon plate tectonics to studies of the shape of the Earth, and he was excited about moving on to other planets shaped by other processes. Then, in late 1986, the shuttle struck again. *Mars Observer* had been scheduled for launch in 1990, but after the *Challenger* disaster the risk of the shuttle's schedule slipping convinced NASA officials to delay the launch until the next time the planets were correctly aligned, two years later. Delaying by two years meant that the spacecraft's costs went up, because it was not feasible simply to disband the teams already at work. Savings had to be made and so the two heaviest instruments were dropped. One was the radar.

David Smith was not going to give up. He convinced NASA to put $10 million on the table to produce a replacement instrument and, having looked at a couple of radars, decided to use a new, much less tested technology, one that bounced laser light off the surface instead of radio waves. People in Smith's group at Goddard were already working on such an altimeter for the proposed Lunar Observer; a modified version became a relatively cheap altimeter for the *Mars Observer*. There were risks involved—no laser system had ever survived in space remotely as long as this one would have to—and the development was a little hairy in places. But they got the instrument finished on time and in budget. That was more than could be said for the rest of the mission. Partly due to the delays, *Mars Observer*'s costs rocketed—the notional later Observers were canceled as a result. Then it was decided to launch on an expendable rocket rather than a shuttle, adding yet more to the expense.* Then a hurricane hit the rocket while it was on the pad at Canaveral. Finally, on 25 September 1992, with the Mars Observer Laser Altimeter (MOLA) safely on board, *Mars Observer* got off the ground. And eleven months later, having been told to pressurize its fuel tanks in preparation for going into orbit around Mars, the spacecraft fell silent, never to be heard from again. It is more or less universally assumed to have exploded.

It was a terrible blow. Back when Mars missions were sent out two at a time, losing one was okay; *Mariner 3* was lost, part of

*Although a shuttle launch costs a lot of money, those costs are not typically borne by any spacecraft along for the ride, but the cost of a one-off rocket is billed to the mission that it launches.

Mariner 7 exploded, *Mariner 8* was lost, but *Mariners 4, 6,* and *9* did just fine. *Mars Observer*, though, was a singleton and the designers of its nine scientific instruments were bereft. Smith told me that while imagining ways in which the MOLA instrument itself might fail had come all too easily to him, he'd never imagined the whole spacecraft being lost. NASA's administrator though—a bullying, obstreperous but undeniably dynamic and often perceptive man named Dan Goldin—decided the loss was an opportunity. Goldin was sick of being responsible for the sort of space program that launched only a couple of planetary spacecraft every decade, and was determined to find ways of sending out more missions—"faster, better, cheaper" missions, as he delighted in calling them. The first faster, better, cheaper program, called Discovery, was to send spacecraft all over the solar system. Indeed, the second Discovery mission, due to take off in late 1996, was a Mars lander—*Mars Pathfinder.* (*Mars Pathfinder* was actually conceived before the Discovery program; as its name implies, it was meant to be the first in a series of simple landers. The series of simple landers was canceled and *Pathfinder,* like *Mars Observer,* became a one-off,* slotted into the Discovery program for more or less purely political reasons.) Goldin and his advisers at NASA headquarters decided that a second line of faster-better-cheaper spacecraft should be devoted to Mars. In order to spur new thinking and greater efficiency, the size of the spacecraft and the budgets in this Mars Surveyor program were to be tightly constrained.

The first of the missions was *Mars Global Surveyor (MGS)* and it has proved massively successful. Launched in November 1996, it arrived at Mars a few months after *Mars Pathfinder*'s landing on 4 July 1997. *MGS* carried copies of five of *Mars Observer*'s instruments, for the most part cobbled together out of spare parts. Soon after arriving it started a long series of passes through the thin upper atmosphere, a way of losing energy to make its orbit shorter and

*This sole-survivor-of-an-imagined-series motif is a common one in the history of NASA; as individual missions grow costly, their proposed successors are canceled. The Planetary Explorer "program" of the 1970s ended up being a single mission. So did the Mariner Mark IIs conceived in the 1980s.

more circular. This technique, "aerobraking," was new and somewhat risky. In the old days before faster-better-cheaper, changing orbits was something you did with engines, not drag. But drag is free and engines cost money.

In the end this aerobraking took a lot longer than anticipated: Most of the atmospheric drag was felt by *MGS*'s solar panels and the arm holding one of these panels turned out to have a flaw in it. The aerobraking sequence was modified so that the spacecraft dipped into the atmosphere even less than had been planned, the force exerted on it ending up as less than three newtons—about the force it takes to lift a Big Mac hamburger. This slowed things down and it was not until early 1999 that *MGS* reached its final orbit, circling the planet every two hours or so, about 250 miles above the surface. The instruments now got down to business. The infrared spectrometer scanned the surface to see what minerals were present, and where. The camera, capable of picking out features just a couple of yards across, started adding long, thin tracks of extraordinary and frequently confusing new details to the coarser pictures of the *Mariner*s and *Vikings*. And MOLA's laser gently zapped the surface beneath the spacecraft ten times a second. The laser beam would illuminate a patch of Mars about 180 yards across and the altimeter's clock would measure the time it took the light to get there and bounce back (less than three-thousandths of a second). The exact length of time revealed how high up the spacecraft was. Combine that altitude with tracking data showing where the spacecraft was—the tracking on *MGS* was exquisite—and you get a point in a global altimetry database. By the middle of April MOLA had produced almost twenty-seven million such altitude measurements. For the most part they were precise to within less than a yard, which means that two nearby spots that seemed to have the same altitude would in reality be no more than three feet different in elevation. The overall accuracy with which the MOLA measurements determined the global shape of Mars was about twenty-five feet.

A year after *MGS* reached its final orbit, in March 2000, planetary scientists from all over America and much of the rest of the world gathered in Houston for the Lunar and Planetary Science Conference, just as they have done every year since 1970, when the first such conference pored over studies of the first samples returned from the moon. For a week, the Johnson Space Center's recreation building was turned over to them, and its basketball courts rang to the announcement of more and more news from Mars. At least a hundred papers on Mars were presented, most of them informed by *MGS* data in one way or another. To many of those attending, Mars seemed to be changing before their eyes. *MGS* measurements were discovering new features and forcing the reinterpretation of old ones. The idea that there had once been an ocean on Mars was starting to gain serious respectability. So was the idea that, far from having been geologically dead for billions of years, Mars was in fact still active. The old familiar face of the planet was taking on a youthful cast in the new light. The scientists were as reinvigorated as their planet.

But if the Houston meeting was full of scientific promise, there was also a fair share of institutional foreboding. Condolences were offered to the people who would have been presenting the first data from the Mars Surveyor program's 1999 missions, *Mars Polar Lander* or *Mars Climate Orbiter*, had it not been for their accidents. Some of these unfortunates—the ones who had worked on an instrument designed to analyze the way the Martian atmosphere changes with altitude—had watched their instrument burn up not once but twice, first on *Mars Observer*, then again on *Mars Climate Orbiter*. The reports of a whole slew of investigative committees on the previous year's disasters were due out in the next few weeks and everyone knew that they would make sad, infuriating reading. While *Mars Global Surveyor* was a wonder, the program it had spearheaded was a disaster.

On a phenomenally wet Tuesday evening, on a set of couches in the foyer of a building on the University of Houston's Clear Lake campus, Carl Pilcher, the man responsible for solar system science at NASA headquarters, discussed the situation with various worried and disaffected scientists. He more or less confirmed that the next Mars lander, one that shared the design of *Mars Polar Lander* and was due to be sent off in 2001, was being canceled. He accepted that the constraints that had been put on the program had proved too tough—that in the effort to force JPL to make the Mars Surveyor program faster-better-cheaper, mistakes had been made both at the lab and at NASA headquarters. He accepted that faster-better-cheaper had meant that the scientists had worked themselves to the bone and encouraged everyone there to help NASA get it right next time round. When he'd finished it was clear that the Surveyor program as it had been talked about just a few months ago, with plans for missions in 2003 and 2005 that would not just study Mars *in situ* but send samples of its surface back to the Earth, was over, and that, as yet, there was nothing to replace it. With one exception—a small orbiter that would carry the last of the *Mars Observer* instruments to their objective in 2001—the future of Mars exploration was, yet again, a blank.

But Mars itself was not. Just across the aisle from Pilcher's attempt to share the pain of his bruised community was a special presentation by the MOLA team. As befits a geophysical instrument, MOLA is in the numbers game. If you put enough numbers together, though, you can get a pretty good picture. The MOLA team had taken their data set, arranged it on a Mercator projection, and printed it out as a map. The first version of this map, published in the journal *Science* the summer before, had been impressive. Garishly colorful, it had shown so much detail in its crater rims and mountain tops that many looking at it had assumed it was a colorful overlay superimposed on some sort of photomosaic or airbrushed map. But every last bit of the picture came from the MOLA data set, from simple measurements of the time it took for a pulse of laser light to reach the surface of Mars and bounce back to *MGS*.

By the time of the Houston conference the map had been much improved. *MGS* had been in its proper orbit for more than an Earth year (though only just half of a Mars year, each of which lasts 687 Earth days, or 669.6 Mars days). More data had been added and very large printers had been used to blow the image up far beyond the scale of a scientific paper. The version in the University of Houston foyer was about six and a half feet long and five feet high. It would have been eye-catching even if you didn't know what it was. If you did know, it was little short of a miracle. Here were real data, as hard and scientific as you could wish, woven into the image of a planet. It was not a realistic image. The altitude data were color coded, so that the terrain ranged from blue in the lowlands through green to yellow to red to white. Hellas, the deep basin in the south, looked out like a baleful violet eye; the rise of Tharsis, its three great volcanoes snowy white, was ringed with burning red. Faint features were enhanced by computer filtering, just as they had been in the *Mariner 9* photographs, to exaggerate details. Shaded relief had been added, not by skilled artists, but by a computer program first developed for charts of the ocean floor. It did a pretty impressive job—while still suggesting, as all such shading does, that the planet knew no night and that the sun was somewhere over the north pole. No, the map was not realistic. But to the people who walked by, and stopped, and stared, it was very real.

I watched for an hour or so as almost every scientist with any interest in Mars passing by on the way to or from the poster presentations elsewhere in the building stopped to stare at the MOLA map. They talked to each other; they pointed out features. They got close and squinted, then stepped back to take it all in. They enthused and gestured, and then fell silent and just stared. Peter Smith, designer and operator of *Mars Pathfinder*'s camera and, in his youth, a photographer with serious artistic ambitions, said it was the most incredible picture he'd ever seen. Baerbel Lucchitta, a striking, stately geologist who has been at the USGS in Flagstaff since the early 1970s, traced her favorite features with a little girl's grin. When people finally walked away, their eyes and minds full, they couldn't help but look back over their shoulders to get just one more glimpse. Here was a map that was most definitely being treated as an icon.

And David Smith just stood by his team's creation and beamed. Other people on the MOLA team have told me that they always expected to put together such a picture of the planet, but Smith says he had had no idea the endless stream of data points would add up to such a striking visual statement. When I'd visited him in his office the year before, when the largest printed version of the map had been about fourteen inches across, we'd looked at the data laid out numerically in vast spreadsheets. Though Smith had been keen to have the biggest possible version of the map printed for the Houston meeting, he'd not actually seen the resultant poster before that Tuesday evening. He was looking at it—and showing it off—for the first time, the joy of it all over his face. Across the aisle from the MOLA map, Carl Pilcher was explaining that an era of exploration that had seemed to be just beginning was coming to an end. But Smith just kept talking and smiling and looking with pride at his map. From time to time he'd touch it, running his hand lightly across the smooth blue of the planet's northern lowlands. As though he could feel the onset of the higher plains to the south. As though the craters might scratch his fingertips.

Two hundred million miles away, an instrument he had argued for and cajoled into being and thought about every day for more than a decade was illuminating the surface of an alien planet ten times every second. And in the rain-soaked Houston suburbs David Smith was stroking the face of Mars, a picture of delight.

Part 2 – Histories

When the investigator, having under consideration a fact or group of facts whose origin or cause is unknown, seeks to discover their origin, his first step is to make a guess.

—Grove Karl Gilbert, "The Origin of Hypotheses"

Meteor Crater

"Craters? Why didn't we think of craters?"

—Isaac Asimov to Frederik Pohl,
on first seeing the images of
Mars from *Mariner 4*

If you care for impressive and beguiling landscapes, Flagstaff, Arizona, has a lot to recommend it. The San Francisco Peaks—remnants of a shattered volcano similar in scale to Mount St. Helens—loom over a town scarcely a hundred years old, wrapped in the forests that attracted its founders. To the south the beautiful canyons of Sedona, carved into the rocks of the Colorado Plateau by water draining from beneath the forests; to the east the spectacular Painted Desert; to the north the Grand Canyon itself, more than a billion years deep, more gazed at and photographed than any other hole the world has to offer. Around the San Francisco Peaks sit lower cinder cones like giant black molehills, weirdly fresh. Some are intact, some thoroughly quarried; their ash grits the roads in winter. One of them, remarkably, is in the process of being turned into a vast meditation on earth and sky, light and stone, by the artist James Turrell, earthmovers his chisels.

If the land is wonderful, so is the sky, which seems to expand in sympathy with the majesty below. The air is dry, clean, a little thin—just the sort of place astronomers like to set up shop. Above the town, amid the ponderosa pines of Mars Hill, sits the telescope through which Percival Lowell imagined the landscapes of Mars. You can go up and have a look through it, if you like; at the right time of year you'll be able to see Mars floating in the eyepiece just as he did, blotchy but beckoning. During my most recent visit it was the wrong time of year, with Mars best seen at about five in the morning, long after Lowell Observatory has closed itself to tourists. But even watched from a motel car park in the predawn glow, Mars seemed closer in Flagstaff than it does in most places, shining clear and bright and true.

To appreciate land and sky together, drive about half an hour east of Flagstaff. A quarter of an hour beyond the line in the landscape where the Coconino forest responds to some subtle cue of altitude or precipitation and gives way to the Painted Desert, you'll find what used to be called Coon Butte. From a distance it looks not unlike the flattened mesas that sit farther off behind it, except for the fact that its heights are a little more crenellated. As you come closer, though, you begin to get the feeling that it is something quite different: smaller, lower, subtly different in form and nature. Rather than sitting on top of the desert like the low flat hills to the south, or puncturing it like the cinder cones behind you, Coon Butte seems to be a bending of the plateau itself, a twisting of the land toward the sky. And so it is.

Coon Butte is one of the places where the sciences of astronomy and geology meet. It marks the spot where, fifty thousand years ago, a very small asteroid's orbit around the sun was cut short by the surface of the Earth. Most asteroids are made of stone friable enough that small ones will explode high above the Earth's surface, shattered by the shock of being slowed by the atmosphere. The fifty-yard asteroid that struck the Painted Desert that day was made of sterner stuff: iron. It pierced the atmosphere intact and plowed on into the planet. Only after it had punched a hole through the surface of the desert did shock waves tear it apart in an underground explosion a thousand times more energetic than that of the Hiroshima bomb, throwing millions of tons of the plateau's rocks back into the sky. The strata of rock surrounding the impact were bent upward, raising the surface of the desert in a ring and forming a sharp upturned rim to the crater. Some boulders were thrown half the distance back to Flagstaff; within three miles the desert was covered with a thick blanket of debris. The hole left behind was 750 feet deep and three-quarters of a mile across, excavated in seconds. The rim of raised rock stood 200 feet or so above the surrounding desert. After fifty thousand years, erosion has smoothed it down to 150 feet.

Meteor Crater, as it is now called, is an impressive sight. By the time you reach the observation area on the north side of its rim—the only part to which the public normally has access—you have driven at least five miles out of your way, you have paid for a ticket, you have walked past a gift shop and a well thought-out visitor's center; you know what to expect. Even so, to come across this sudden theater of steep relief in an otherwise flat desert takes you aback. It is a big, dramatic hole, its base smoothed by the dried-out bed of a little lake, the strata of raw bedrock poking out of its sides like piers in an arena to seat a million.

At the same time, by the standards of truly dramatic valleys, canyons, and volcanoes—standards the Arizona landscape requires all its tourist features to measure up to—Meteor Crater is not really so terribly large. Its sides are steep and deep, to be sure: If St. Paul's Cathedral were built at the bottom, the great golden cross on top of the dome would be well below your eye level. But the depths are enclosed in a way that almost belittles them. Craters are the most revealing of landscapes; from the rim you can quickly take in all there is to see. And this is not that large a crater. The rim is only two and a half miles around. You could walk around it in a couple of hours (the walking is quite hard, for the rim is not regular); your eye runs around it automatically, limiting its scope in the process. Anything you can grasp this easily cannot give a sense of true enormity.

The most striking effect is not to look down from the rim into the crater's depths, but rather to look straight across. To the south, the circle of the crater's rim and that of the farther horizon lie one upon the other, tangent arcs. Turn your head slowly—pan like a camera—and they become detached. The rim falls away from the true horizon; it twists into the middle distance, banking toward you as the true horizon keeps its distance, becoming a feature within the landscape rather than a limit at the edge of it. Eventually it ends up under your feet, a rampart of rubble dividing the bowl enclosed within from the great desert plain outside. And yet the rim still feels linked to the horizon itself. The great circle of the planet and the ring of the rim seem aspects of the same thing; the great void below echoes the great vault above. The effect has something in common with the old cliché of a straight road, a flat plain, and a vanishing point on the horizon. But here there are no points and lines and directions: just circles turning in on themselves over 360°. This sense of a world arranged in nested circles may be something nothing else can offer as well as a deep astronomical impact with a well-preserved rim. And on the Earth, there are no other impact craters with rims as well preserved as Meteor Crater's.

On Mars, by way of contrast, there may be a quarter of a million impact craters the size of Meteor Crater. And there are craters of all other sizes, too. There are great impact basins large enough to put the European Union into; there are craters small enough to use for tennis courts. There are craters that overlap like the circles of an Olympic flag. There are craters on the rims of bigger craters. There are craters within craters within craters. Some are as young as Meteor Crater itself, or younger. Some are more than eighty thousand times older, landscapes more ancient than anything on the Earth's shifting surface except a few tiny zircon crystals preserved by chance.

And those are just the ones you can see in the airbrush maps and the *Viking* pictures; the ones with clear, well-defined rims. One of the discoveries made with the data from *Mars Global Surveyor*'s MOLA altimeter was that there are hidden craters, too, craters yet more ancient than the visible ones, if only by a little. The MOLA team has developed all sorts of ways of using brightness and color cues to bring out different aspects of their vast data set. One of their best tricks is a way of looking at the planet slice by slice on a computer screen. The spectrum of colors that allows the eye to understand what it is seeing is concentrated into a thin range of altitudes—less than a thousand feet, perhaps—with all lower places darkly blue, all higher grimly purple; the highlighted range can be moved up or down at will. Look at a mountain this way and you will see a circular band of rainbow with a dark center. Toggle the highlighted range upward and the noose of light will tighten to a solid disc at the summit; lower it and the ring of color will expand slowly until it smears itself out across the plains at the mountain's base.

Run this magical palette over the surface of Mars and crater rims will stand proud as thin, hollow crowns. But rimless craters can be found too: solid circles of equal altitude. These are old, eroded craters, craters the unaided eye would never pick up. These shadow craters can be quite big: One of the first to be discovered this way was about 280 miles across, giving it an area about the same as Michigan's or England's, and definitely putting it in the first division of Martian craters. And they are quite numerous. In the summer of 1999, seventy flat circles of various sizes that looked like ancient impact scars were discovered by one high school student doing an internship with the MOLA geology team.

Discoveries on such a scale mark a peculiarly auspicious beginning to a scientific career. But if the intern was spectacularly successful in how she did her job, she was not particularly distinctive in the job she was doing. Almost every geologist who looks to the skies as well as the Earth starts off counting craters; most will still be doing so, now and then, decades later. It is the way that astrogeologists measure time. On the Earth, geological time is measured in layers; layering is history and depth is age, as a drive to any edge of the Colorado Plateau will demonstrate. Stratification, though, like embonpoint, is best seen in profile; on planets looked on only from above, the study of strata is geometrically challenging, to put it mildly. But craters, too, are the testaments of time; like sediment on a sea floor, they accumulate over the years. Most planets with rocky surfaces are amply supplied with craters: The Earth, endlessly reinventing its surface through erosion and plate tectonics, is the great exception. Reading the record of craters has made sense of the geology of the moon, has revealed global cataclysms responsible for remaking the surface of Venus, and has provided, at least in outline, the history of the Martian surface from the most recent sharp-edged scar to the most ancient rimless basin.

It was through Meteor Crater that people first learned how to read such records. It gave them what geologists most need: an analogue through which to understand processes not yet understood in any other way. Analogy sits at the heart of geology; it has long linked the past to the present, and now serves to tie the earthly to the alien. Meteor Crater allowed geologists to understand impacts, and its nested horizons became the door to other worlds and other times. Its role in understanding was not just theoretical. In the 1960s Meteor Crater was one of the sites chosen to train the only men from Earth ever to walk anywhere else; strain your eyes and by the lake bed at the bottom you can see the statue of an Apollo astronaut that commemorates them. From his point of view there is no double horizon; beyond his little bowl of a world there is just the great urgent vault of the sky.

"A Little Daft on the Subject of the Moon"

> We pride ourselves upon being men of the world, forgetting that this is but objectionable singularity unless we are, in some wise, men of more worlds than one.
>
> —Percival Lowell, *Mars*

The story of how Meteor Crater came to be understood as the best-preserved earthly exemplar of the ancient landscapes of Mars comes in two parts. In the first part a great geologist got it wrong, but in doing so showed how geology could, in principle, tackle subjects beyond the Earth. In the second part a great geologist got it right and used his insights to turn the geological mapping of other planets, including Mars, into a practical concern. Both men were geologists with the U.S. Geological Survey and both were filled with the romance of the American West—a romance that both science and the popular imagination have projected onto Mars for more than a century.

Grove Karl Gilbert was one of the happy generation of American geologists who, in the second part of the nineteenth century, took their impressive beards and intellects to every corner of the American West. They were part of a worldwide phenomenon that the historian William Goetzmann has called the second age of exploration—the period between Cook's voyages in the late eighteenth century and Amundsen's trek to the South Pole in the early twentieth during which Europeans moved beyond the coastlines of other continents and across their hearts. The centers of Africa, Asia, and Australia were all explored at this time.

In the American West Gilbert and his peers—John Wesley Powell, Clarence King, Clarence Dutton, William Davis—encountered a world that spoke to them of the archaic and at the same time cried out for the modern, an awe-inspiring natural world that could only be opened to civilization through the technologies of electricity, irrigation, and the railroad. Its forbidding landscapes—often wonderfully captured by the artists and photographers who accompanied the various expeditions—were utterly unlike those the scientists were familiar with back east; plateaus dissected by massive erosion, the strangely faulted terrains of the basin and range province, all manner of volcanic dramas. The explorers measured the landscapes, mapped them, and developed new language to describe them: "laccolith," "isostasy," "gradation." Dutton, in particular, was a literary gent (as well as a soldier, a chemist, and a theology school dropout) who styled himself "omnibiblical"; his writing overflows with energy. His descriptions were evocative, grandiose, and sometimes extremely funny, continuously aware of his audience back east and the novelty he was bringing to it. In his memoir of the Grand Canyon, the Survey's first publication, he wrote in self-justification:

> I have in many places departed from the severe ascetic style which has become conventional in scientific monographs. Perhaps an apology is called for. Under ordinary circumstances the ascetic discipline is necessary. Give the imagination an inch and it is apt to take an ell, and the fundamental requirement of the scientific method—accuracy of statement—is imperiled. But in the Grand Canyon district there is no such danger. The stimulants which are demoralizing elsewhere are necessary here to exalt the mind sufficiently to comprehend the sublimity of the subjects. Their sublimity has in fact been hitherto underrated. Great as is the fame of the Grand Canyon of the Colorado, the half remains to be told.

Dutton, Gilbert, and their peers did not just find new language with which to express themselves; they came up with new theories about how the Earth might work, theories that allowed for far greater violence and more sudden novelty than the sedate forms of geology practiced by their European forebears and contemporaries. It was the need to explain the landscapes of the West, and the mineral wealth they might hold, that led to American geology becoming a nationally distinct enterprise quicker than any of the country's other sciences.

By the time he came to Meteor Crater in 1891 Gilbert had spent twenty years, as he put it, "aboard the occidental mule," trying to understand the processes that had shaped the landscapes around him. He was a precise man, mathematically orientated, but he also had a zest for the experiences that would help him explain how the landscapes he carefully measured had come to be. It was, he wrote, "the natural and legitimate ambition of a properly constituted geologist to see a glacier, witness an eruption and feel an earthquake." When he achieved the last of those ambitions in 1906, it was with "unalloyed pleasure"; awakened by the shocks of the San Francisco earthquake, he set to timing them and measuring their direction. He brought the same precision to his other work, closely harnessed to a love of physical and mechanical analogy.

His love for the orderly and mathematically tractable led him to study the stars as well as the Earth. Traveling down the Grand Canyon on one of the first expeditions to do so, he had made a point of observing Venus from its depths. He was by his own admission "a little daft on the subject of the moon," and in Washington, D.C., he made use of the Naval Observatory's telescopes to observe it in detail, prompting ridicule from congressmen who affected to think that if distinguished members of the U.S. Geological Survey had nothing better to do than look at the heavens, the Survey should clearly be disbanded, its earthly work complete. In Gilbert's thought, though, geology and astronomy belonged together; together they could explain not just rocks but entire planets.

In the summer of 1891 a Dr. Foote reported to the American Association for the Advancement of Science that he had found significant amounts of meteoritic iron at Canyon Diablo in the Painted Desert, near the crater at Coon Butte. Gilbert was intrigued. He thought matter falling from the sky might shed light on what he saw as one of the great planetary problems: why the Earth's crust is systematically denser in ocean basins than under continents. Gilbert thought this heterogeneity might be due to the fact that the Earth had been assembled from smaller objects, which later theorists would call planetesimals: dense crust marked the contributions of dense planetesimals. Gilbert wondered whether the large crater in this field of meteoritic iron marked the spot where a "small star" had been "added to the earth" relatively recently. Always ready to head west when possible, he arrived at Meteor Crater that October.

Gilbert saw two possible types of explanation for the crater: It could have been formed by something coming in—an impact—or by something coming out—a volcanic explosion. The best argument for a falling star was the meteoritic iron littering the surrounding desert. Gilbert calculated the odds of a crater forming in such a dense meteor field purely by chance as 800 to 1. If the crater had been clearly volcanic, then this might not have mattered. But though there were volcanoes nearby, the crater's walls and floors were sedimentary rock, the same strata of sandstone and limestone from which the rest of the Colorado Plateau is built.

In a typically methodical manner Gilbert set out to test the alternatives through their implications. If there were a "star" buried beneath the crater somewhere, then like Archimedes in his bath it would have displaced material that was there before. If so, there would be more material in the crater's raised rim and its surrounding blanket of ejecta than was needed to refill the crater itself. But when, through painstaking surveying, Gilbert and his assistants compared the volume of the crater's cavity with the volume of the rock that had been excavated in the catastrophe, they found that if the rim and ejecta were put back into the crater they would almost exactly fill it up; thus there was no evidence for the bulk of an added meteor below the crater floor. What was more, if a large iron meteorite did lie buried there it should have had a quite discernible magnetic field. But no such field was found. So Gilbert decided the crater had been formed by an explosion of steam, set off when deep volcanic activity had penetrated a subterranean aquifer; he placed Coon Butte in the family of anomalous volcanic craters called "maars" (no relation). This hypothesis sat well with the natural inclination of the area's uneducated shepherds: that the crater looked as though it had been formed by something exploding out of the Earth, not by something falling into it.

Disappointed as he may have been—a maar is an interesting

thing, but hardly a star—Gilbert still put his observations to good use. In his 1895 address as president of the Geological Society of Washington, published as "On the Origin of Hypotheses," he presented the story of Coon Butte as a sort of moral fable on the correct way of approaching geology. To explain a novel feature, the geologist should first reason by analogy: What sort of thing is it like? The analogy might seem a distant one—a gaping crater in a desert is not very like the "raindrop falling on soft ooze" to which Gilbert compared Coon Butte—but that need not matter. What matters is that there be a number of analogies, that they have different physical implications, and that those implications then be tested. This was Gilbert's highly influential encapsulation of what was becoming the pragmatic cornerstone of geological science in America: a method of "multiple working hypotheses" in which contradictory explanations were to be entertained simultaneously.

One of the disappointments for Gilbert in finding Meteor Crater to have been produced from within and not without was that he had hoped to use it as an analogy with which to bolster his theories about the moon. Everyone who wrote on the moon explained it by analogy to the Earth; the problem lay in choosing the right analogy. In 1874 James Carpenter, a Greenwich astronomer, and James Nasmyth, an engineer whose father had been a landscape artist and whose own pictures of the moon had caught the eye of Prince Albert, published a wonderful illustrated book called *The Moon: Considered as a Planet, a World, and a Satellite.** Inside, spectacular photographs and prints of the moon are compared with similarly lit photographs of a range of other objects—an old man's wrinkled hand, a desiccated apple, a cracked sphere of glass. The idea is to teach the reader's eye new ways of seeing the moon and give his mind new analogies by which to understand it. (Their influence was

*In *View Finder*, the poet and author William Fox recalls finding one of Nasmyth's prints on the wall of Mark Klett, an Arizona photographer with a taste for nineteenth-century geological records: "Some photographs automatically elicit covetousness and this is one . . . Science remains a highly romanticized part of our social endeavors; combined with art, it makes for irresistible artifacts, which is also part of the reason we so love maps."

long lasting. Lowell used the desiccated apple in his books on Mars to demonstrate what happens when a planet dries up; the first post-Mariner textbook on Martian geology has very Nasmyth-like cracked glass spheres in it to demonstrate stress patterns.)

To make their case for the volcanic origin of the moon's craters, Nasmyth and Carpenter created a scale model of what Vesuvius and the bay of Naples must look like from above and compared it with similar models of the lunar surface. Other lunar analogies on offer suggested that the dark expanses of the moon called "seas" were in fact made of ice, or that they were the dried beds of seas now vanished. Charles Babbage, the pioneer of mechanical computing, elaborated on this idea with the notion that craters in these dried seas were in fact coral atolls like those studied by Darwin.

Gilbert rejected all these analogies, seeing the craters and larger basins and "seas" as the marks left by planetesimals. His idea was that once the Earth had been ringed by planetesimals—much as Saturn is ringed today—and that these had then coalesced into the moon; the last ones in had left the surface scarred. Lacking a natural earthly analogue for such cratering, Gilbert experimented with crater making himself, firing various projectiles into clay; he called the hobby "his knitting" and found the results satisfactorily lunar. To those who objected to a geologist trespassing in the realms of astronomy, he defended his speculations in terms that could serve as the credo for astrogeology to this day, "The problem is largely a problem of the interpretation of form, and is therefore not inappropriate to one who has given much thought to the origin of terrestrial topography."

Gene Shoemaker's thinking on terrestrial topography, which would find application in the interpretation of form on the moon and beyond, took place in large part on the Colorado Plateau that Gilbert had known so well (indeed, he had given it its name). In 1948, twenty years old and with a Caltech degree in geology already behind him, Shoemaker joined the U.S. Geological Survey and found himself working in southern Colorado. He discovered that he loved the landscapes of the Southwest. He loved the pines, he loved the open spaces, and he loved the great, vaulting skies. He stared up at the desert moon with wonder.

In the field, he did not have much contact with the rest of the world. But he did get the Caltech alumni newspaper, which revealed that experiments with captured V2 rockets elsewhere in New Mexico were reaching the very edge of the atmosphere. It was a revelation. "Why, we're going to explore space," he later remembered thinking, "and I want to be part of it! The moon is made of rock, so geologists are the logical ones to go there—me, for example."

Shoemaker kept his wild dream to himself—a decade before Sputnik there was little call for space-age geology. The atomic age, though, needed geologists badly. Cold War strategy required that America develop reliable domestic sources of uranium, and the Colorado Plateau was thought likely to hold the reserves required. So in his first years with the USGS Shoemaker joined in the last great American mining boom; at the same time he started work on a Ph.D. at Princeton and got married. He crisscrossed the Colorado Plateau from site to site, "half man, half jeep," according to his wife, Carolyn, who often accompanied him. It wasn't normal for geologists' wives to come along on field trips, but the Shoemakers didn't care. When they had children, the children came too.

All the while, Shoemaker kept thinking about the moon. He read everything there was to read on the subject, including Grove Karl Gilbert; he tailored his fieldwork to suit his extraterrestrial interests. It was this that led him to map diatremes in the Painted Desert's Hopi Buttes. Diatremes are volcanic features, chimneys of magma that rise to the surface causing explosions that throw out a lot of normally well-buried rock and comparatively little lava; they can create the low-lying craters called maars to whose number Gilbert had added Meteor Crater. As a uranium prospector, Shoemaker was interested in diatremes because the rocks they threw out when they cleared their throats might be from uranium-bearing strata. As a would-be lunar geologist, he was interested in them because their associated craters often occur in families laid out along a straight line; many lunar craters show a similar linear tendency. Diatremes might thus be analogies by which to understand some forms of volcanism on the moon.

Shoemaker caught his first fleeting glimpse of Meteor Crater in the late summer of 1952. One afternoon, driving past the town of

Winslow, he convinced his wife and a colleague that the site to which the great Grove Karl Gilbert had devoted his time might be worth a look. They didn't have the entrance fee required for the public viewing platform on the north edge of the crater, so they had to take an indirect approach via a dirt track and then scramble up the rim. By the time they got to the top, the sun was setting and most of the great bowl was already in twilight. They stayed only a few minutes and Shoemaker saw nothing to contradict Gilbert's assessment. In a few years, though, he would. A landscape that was only then being brought into existence gave Shoemaker the analogy he needed to understand Meteor Crater.

By the mid-1950s the American uranium rush had uncovered deposits of the stuff large enough to meet any plausible need. Plutonium, though, was another matter. Natural plutonium exists in only the tiniest of quantities, on the order of a kilogram per planet or so. If you want to make bombs of plutonium—which has a number of advantages over the use of uranium—you first have to produce the stuff. This is normally done with the help of a nuclear reactor that breeds plutonium from uranium. However, in the mid-1950s a quicker alternative started to be discussed: simply wrapping nuclear weapons in now plentiful uranium and setting them off somewhere where their blasts would be contained. Neutrons given off in the explosion would turn some of the mantling uranium into plutonium, which could then be scraped up and put into more bombs.

A necessary part of this alarming scheme was finding places where you could set off nuclear bombs without the debris being scattered to kingdom come. One possibility was salt caverns; since Shoemaker's Princeton work dealt with salt structures in southern Utah, he was called on to look into the question. As a result, he saw what few other geologists had had the chance to see: a pair of craters at the Atomic Energy Commission's Nevada test site that had been formed by underground nuclear explosions, craters called Teapot Ess and Jangle U. The shapes of the craters, he learned, were determined by the interplay of various sets of shock waves, some heading out from the bomb, some bouncing back. The nuclear-testing fraternity had a keen understanding of shock waves; a correct calculation of the behavior of the shock waves inside their bombs was the key to getting them to go off properly in the first place. So Shoemaker learned of the power of shock waves both from the physicists and from the craters themselves, their edges deformed by enormous pressures, the sand around them fused to glass.

When Shoemaker went back to Meteor Crater in 1957 it was not directly because of his new experience in matters nuclear (he advised the bomb makers, incidentally, that keeping explosions contained underground was not going to work). But that new experience changed the way he saw the crater; now it looked like the aftermath of something like a nuclear explosion, something formed by shock waves in a matter of seconds. He set about making a systematic study—a process that required that he make friends with its owners, the Barringer family. In the early decades of the twentieth century Daniel Barringer, a lawyer and mining engineer, spent a lot of time trying to convince people that the crater had been formed by an impact, and a lot of money trying to mine its floor for the huge and valuable lump of pure iron he expected to find there. His failure to convince the world that it was even worth looking for such iron was in part due to the fact that no less an authority than G. K. Gilbert had disagreed with the idea. Barringer's heirs had inherited both a largely useless crater and a dislike of geologists from the Survey. Shoemaker, though, became their friend and ally, in part because he was introduced to them by their old schoolmaster.

Shoemaker's work at the crater in the late 1950s both vindicated Daniel Barringer's insight and revealed his dream to have been illusory. Shoemaker found places where heat had turned sandstones into glass; he mapped inverted strata on the crater's rim where the lip had been folded back on itself just as the rim of Jangle U had. Most crucially, he and a colleague found that the sandstones in the crater contained coesite, a very dense mineral created when quartz is subjected to extreme pressures. Only transient shock waves of immense power could create such pressures on the scale required; coesite was frozen smoke from the impact's gun. Meteor Crater was indeed caused by an impact, as Barringer had thought. But the rebounding shock waves that formed the crater had completely exploded the incoming lump of iron he had hoped to profit from. As Gilbert's volumetric assessments had shown, there was no extraterrestrial mother lode beneath the floor.

While Shoemaker worked to forge this causal link between heavens and Earth, a matching connection was being set up in the opposite direction; the first earthly objects were being sent beyond the planet. Exploration of the moon suddenly seemed a practical possibility and Shoemaker was keen that the USGS should grasp it—just as Gilbert would have been. The Survey was keen, too. It had expanded a lot during the search for uranium, and now that that search was over, and the Atomic Energy Commission had withdrawn funding, it was left with more geologists than it had money or jobs for. Shoemaker's astrogeological dreams might take up the slack—if funding could be found.

The obvious place to look was the newly formed National Aeronautics and Space Administration, but NASA was cagey about handing over the money. It was the huge success of Shoemaker's work on impacts, and in particular his demonstration, in Germany, that the presence of coesite could be used to show that much larger, more highly eroded features than Meteor Crater shared its extraterrestrial origin, that finally got NASA on board. By August 1960 Shoemaker and a handful of others made up a newly formed Astrogeology Study Group, half of them in Washington, D.C., and half in Menlo Park, California. When in 1961 President Kennedy committed America to reaching the moon within a decade, the astrogeologists were well positioned to be part of the adventure.

Shoemaker did not need to wait for the first moon missions to study the processes that shaped it. Now that he understood the process of impact cratering in all its phenomenal violence, he was able to see how it accounted for much of the lunar surface. At the time, the origin of the moon's craters was still a topic of hot debate, with the camps divided between explanations invoking volcanism and those invoking impact. Shoemaker argued that the unearthly features around large lunar craters—bright, far-reaching rays, confused hummocky terrains, smaller craters in lines and clusters—could be explained by the physics of the impact. Near the crater thick blankets of ejecta would smother the surface; farther off, fragments thrown high into the sky would come back to the surface moving fast enough to create new craters of their own, known as secondaries, arranged in lines and clusters.* Ejecta moving almost horizontally across the surface tore holes in earlier features; a torn terrain of such scars, known as the Imbrium Sculpture, surrounds the Imbrium basin.

In 1960, on a trip to JPL that was in part an exploration of employment opportunities, Shoemaker was astonished to see one of the earliest copies of the first moon map drawn by Pat Bridges: a map of Copernicus, one of the youngest and most striking of the moon's craters, and one that he had been studying as he worked out his ideas about ejecta blankets and secondary craters. He got a copy and, as soon as he returned to Menlo Park, began to use it as the basis for his first lunar geological map.

On Earth, geological mapping starts in the field. The geologist wanders the landscape from outcrop to outcrop, identifying the rock type in each one. He assigns the outcrops to various geological units—bodies of rock formed by a single process, or a set of related processes, in a discrete period of time. The result shows which unit is closest to the surface at any given point. It shows the surface as a two-dimensional slice through a three-dimensional structure that itself bears the clear imprint of developments through the fourth dimension of time; a well drafted geological map is one of the densest, most informative forms of illustration known to humanity. A geologist can take such a map and learn as much from it as a detailed, definitive book, one that explains what the geological units in question are, how and when they were created, and in what order, and in what ways they have been twisted, raised up, eroded away, and turned into the complex mishmash presented on the surface today. A good map may represent most of a lifetime's fieldwork.

The fact that he couldn't do any fieldwork on the moon (something that, at that point, he still hoped to rectify) put Shoemaker at huge disadvantage: He couldn't actually sample the rocks that make up the different geological units. He could only go by the lay of the land, ascribing the surface to various different geological units on

*Secondaries, not diatremes, thus account for the moon's lines of craters.

the basis of its appearance. But the map that Shoemaker produced using photographs and Pat Bridges's airbrushed base showed that this could be enough to get good results. Looking at the Copernicus region, Shoemaker was able to distinguish a variety of geological units by their surface textures. There were units of the Imbrian system, a gently rolling surface studded with closely spaced low hills and intervening depressions; there were units of the Procellarian system, dark, smooth, and gently ridged; there were various deposits associated with various craters, including those created at the same time as Copernicus itself—the crater rim, the ejecta, the outlying rays of material thrown out farther, and the crater floor. Studying the places where the units met, it was possible to place them in stratigraphic order, saying which units overlaid which others. On the basis of straightforward superposition—the rule of thumb that the topmost units are the youngest, which outside severely contorted mountain belts and their ancient remains is a pretty good one—Shoemaker put the region's history together. He recognized that a vast impact—the one that had formed the Imbrium basin to the northwest—had covered most of the area with a rolling, lumpy blanket of ejecta. The smooth Procellarian rocks overlying this blanket were from a later period when free-flowing basaltic lavas had filled such basins. He saw that the various later crater deposits were from two subsequent ages, first the Eratosthenian and, finally, the Copernican. Copernican debris overlaid Eratosthenian debris, thus establishing their relative ages. What looked to the untrained eye like a hugely complex mess was in fact a relatively straightforward progression in which each feature had its proper place.*

Shoemaker's lunar mapping studies, eventually published in 1962, proved that unearthly terrains could be subjected to geological analysis, as long as there were pictures to work from and a well-trained eye to work with them. That same year, at the suggestion of his colleague Dan Milton, Shoemaker came up with the idea of moving the Survey's astrogeologists, who at this stage were focused

*Further studies redefined this framework: Lunar geological history is now defined in terms of five periods, the Pre-Nectarian, Nectarian, Imbrian, Eratosthenian, and Copernican.

entirely on the moon, to Flagstaff. There was a fine observatory the Air Force was already using for lunar mapping, and there was a range of geological formations that seemed relevant to the moon—cinder cones, diatremes, and, of course, Meteor Crater itself.

By the time a fair proportion of the astrogeologists had moved out to Flagstaff, though, Shoemaker himself had already moved on, seconded to NASA. Thanks to President Kennedy, by this point America as a whole was going a little daft on the subject of the moon, embarking on a ten-year program aimed at the grandest achievable technological goal imaginable before a single American astronaut had gone into orbit. At NASA, Shoemaker was instrumental in making sure that this magnificent daftness would have a scientific component. As a result of his efforts, astronauts were to become regular visitors to Flagstaff, from where the astrogeologists would take them off on field trips to the volcanoes and Meteor Crater, the Grand Canyon, the San Juan Mountains in Colorado, and anywhere else that might be relevant, anywhere that might help to train their eyes.*

Gene Shoemaker himself might have been one of the astronauts, had it not been for the fact that while learning to pilot a plane in the early 1960s he discovered that he had an adrenal condition called Addison's disease that would disqualify him from flight. So the visionary affectionately mocked by his colleagues as "Dream Moonshaker" had to be content with creating the institutional and scientific framework for lunar geology—both of which would, in the early 1970s, be turned to the study of Mars. The techniques of mapping from photographs that Shoemaker had pioneered, and that he had insisted every astrogeologist become conversant with, would be used to create geological maps that tried to explain, or at least to order, the patchwork world of smooth and cratered plains, volcanoes and valleys and canyons that *Mariner 9* and the *Viking*s

*They didn't do all their studying in America; some went to Lonar, in India, where an impact crater similar in size to Meteor Crater punctuates the flood basalts of the Deccan Traps. Being in basalt, the volcanic rock that makes up the surface of the lunar maria, this crater is in some ways the best analogue to the moon that the Earth has to offer. It also contributed to Gilbert's error over Meteor Crater; knowing that in India there was a very similar crater in a volcanic setting was one of the reasons that he ended up deciding the Arizona crater was volcanic, too.

revealed. And the Colorado Plateau landscapes he loved would provide analogies to explain their formation.

Shoemaker and Lowell are not the only reasons why Arizona is a center of planetary science. Before Shoemaker moved the astrogeology branch of the Survey to Flagstaff, the Dutch astronomer Gerard Kuiper had set up a lunar and planetary science institute at the University of Arizona in Tucson, to take advantage of the nearby Kitt Peak observatory. In between the two of them, a significant planetary science group grew up at Arizona State University (ASU), in Tempe. Thanks to these institutions Arizona boasts the highest per capita proportion of incipient Martians in the world. And this seems entirely fitting, because when the world thinks of Mars, the images it conjures up are, more often than not, images of the American West.

Look at the pictures of Mars on book covers or movie posters, and you'll see little that you couldn't find on tourist brochures from the Flagstaff chamber of commerce. Steep-sided canyons in sedimentary rock, isolated slabs of the same rock in Monument Valley–style mesas, flat deserts painted red. The pictures of Mars in Kim Poor's Novagraphics gallery, the only commercial gallery in the world to specialize in realistic depictions of outer space—and located, far from coincidentally, in Tucson—are often remarkably similar to the Arizona landscapes commemorated in tourist art all over the rest of the city. Only the skies, with their multiple moons and unearthly coloring, show the difference.

There's some truth in these portrayals. While the features of Mars are very unlike those of the American West in scale and history, the planet undoubtedly looks more like Arizona than it does Arkansas or the Ardennes. But physical resemblance is not the whole story; this representation of Mars is ideological, as well. Since the 1960s, America has ceaselessly talked about space as a new frontier, a continuation of the nineteenth century's expansion—and one result of this rhetoric has been an eagerness to see the landscapes of the solar system in terms of the landscapes of the American West. When Neil Armstrong told the watching world that the "stark beauty" surrounding Tranquillity Base looked "like the high desert of the United States," he wasn't just speaking on the basis of personal experience or of the training with which the Survey had equipped him. He was expressing the rhetorical continuity between the pioneers of the space age and the pioneers of the West on which his adventure was founded. One of Kim Poor's own images, painted long ago for a rock album cover and reproduced on T-shirts and postcards ever since, makes the point beautifully. In a desert landscape, saguaro cacti are lifting off on tongues of flame like misshapen rockets. A cliché of the old frontier is transformed into the icon of the new.

The frontier metaphor is applied to space exploration of all sorts, from Apollo to *Star Trek* to the International Space Station. But when applied to Mars it gains a special immediacy. Mars may not be very similar to the American West—but it is similar enough to give the metaphor substance. Well before the space age, it was natural for American imaginations to see Mars in terms of the West: remote, dry, and a test of character. Lowell, sitting in a region where the politics of irrigation carried all before them, formed the planet's indistinct features into a network of canals. In the popular Barsoom stories, Edgar Rice Burroughs catapulted his hero John Carter from the mountains of Arizona, where he faced death at the hands of the savage red man, to the dried sea beds of Mars, where he faced death at the hands of the savage green man. Ray Bradbury's Martian Chronicles added the disinheritance of the natives to the mix, his Martians dying of chickenpox as the earthly invaders played harmonicas around their campfires.* With all that in the background, how could the volcanoes and canyons revealed by *Mariner 9* not have been read in terms of the landscapes of the West?

Given all this, it might seem fitting that such a high proportion of the Earth's Martian geologists live and work in Arizona, with even more just over the border in California. The situation is not without risk; Mars is not the Colorado Plateau, and if you convince yourself

*The analogy is quite specific, as shown in this exchange among earthmen afraid they are being stalked in the story "– and the Moon Be Still as Bright":

" 'Let me ask you a question. How would you feel if you were a Martian and people came to your land and started tearing it up?'

" 'I know exactly how I'd feel,' said Cherokee. 'I've got some Cherokee blood in me. My grandfather told me lots of things about Oklahoma territory. If there's a Martian around, I'm all for him.' "

it is, you will miss much of what it's trying to tell you. Happily, Arizona's astrogeologists understand this danger. Talk to Alfred McEwen at the Lunar and Planetary Laboratory in Tucson about why he came to Arizona and he'll talk about the inspiring landscapes; ask about his work on Mars, though, and you'll learn about similarities to Icelandic lava flows and Indian basalts. Talk to Ron Greeley at ASU in Tempe and you'll hear that the American West is the place to be for a geologist; but when talking about Mars he'll tell you about windblown Australian soils and the ramparts of the Hawaiian volcanoes. Go to the Gilbert building in the little USGS campus perched on a mesa above Flagstaff and you're likely to hear people planning a weekend hike in the Grand Canyon; but talk of Mars focuses on glaciers and permafrost and ice-streams.*

Are these people influenced by the landscapes around them? Certainly. No one with the feeling for landscape that makes a geologist—the feeling that held Shoemaker in its grip from his first fieldwork to the day in 1997 when he died driving to an impact crater in the Australian outback—could live in such a beautiful and geologically fascinating part of the world and not have it work its way into his or her thinking. But they are much more influenced by the approach that led to those landscapes being understood in the first place: that of systematically mapping the terrain and of seeking out useful analogies wherever they may come from; that of formulating multiple working hypotheses and of using your reason and imagination to choose between them. They may live in Lowell's state, but they are Gilbert's grandchildren.

*The Gilbert building houses most of the astrogeologists. The other buildings are called Powell, Thompson, and Dutton, named for the first explorers of the Grand Canyon, which is the other thing that USGS staff at the campus study. The building now under construction is called Shoemaker.

An Antique Land

> A typical fantasyland will display—often initially by means of a prefatory map—a selection, sometimes very full, from a more or less fixed list of landscape ingredients that includes the following features: a continent (or two), one or more inland seas and an ocean (or two), archipelagos, mountains, isolated islands, fjords, steppes, pastures, deserts, forests . . . and realms of ice, edifices and cities, usually ancient, sometimes abandoned.
>
> —John Clute and John Grant, *The Encyclopedia of Fantasy*

At this point—before any more analogies, before any more of the history of their making, before any analysis of their implications—it's time to look in more detail at what the maps of Mars actually show. So here is a primer on the geography of Mars.

Among the first duties of exploration, as Gilbert and Powell and Dutton discovered, is the naming of names. As they saw the surface of Mars come into focus beneath their camera, the *Mariner 9* team gave the things they were seeing informal names—the Spots, the Chandelier, Carl's Creek, Arroyo Murray, the Elephant's Ass, the Inca City, and so on. At its 1973 general meeting in Sydney the relevant committee of the International Astronomical Union devised a rather more erudite and much more systematic nomenclature. Almost two hundred craters were named after people associated with Mars, a tradition that continues to this day: In August 2000 I watched the current incarnation of that same committee, gathered in a sociology classroom at the University of Manchester, ratify the naming of a crater in honor of Carl Sagan close to that of his old colleague Hal Masursky. (At the same time they gave Gene Shoemaker an eternal memorial near the south pole of the moon.) As time has gone on, though, the demand for names has happily far outstripped the supply of dead Mars scholars, and so craters can now also be named after small earthly towns and villages. Even so, only a small fraction of the planet's craters are named: Most are known only by the letters of a map reference. Many have probably never been the subject of a spoken sentence or a written description, some not even of a conscious thought.

To deal with features other than craters, the committee turned to a range of descriptive Latin landscape terms that had been put together by various geologists, mostly at the USGS. Names for specific features were made by pairing one of these terms with a proper name taken from the old albedo maps of Schiaparelli, Lowell, and Antoniadi. Thus a plain would be a planitia if low-lying, like Arcadia Planitia, and a planum if elevated, like Daedalia Planum. A raised feature could be a mons, like Olympus Mons, a tholus (a hill—something like a mons, but smaller), a patera (volcanic-looking but low and sort of squished), a mesalike mensa (often in the plural, mensae, to describe dissected tablelands), or a dorsum (ridge). A narrow, straight trough was a fossa (ditch); a deeper, more pronounced canyon would be called chasma. A string of depressions along a line was a catena (chain); a widespread area where the surface seemed to have collapsed in on itself was a chaos. All these terms were, in time, applied to other planets and moons as well, from Caloris Planitia on Mercury to Kraken Catena on Neptune's moon Triton.

The most evocative class of feature discovered by *Mariner 9*—the winding valleys that appeared to have been made by erosion—got the most poetic family of names. The term introduced for these valleys was vallis, contrasting them with the fossae and chasmae that seemed to have been made by tectonic movements in the planet's crust. Carl Sagan made the inspired suggestion that these valleys be given the names by which Mars itself was known in various different earthly languages, breaking the Greco-Roman traditions of the rest of the procedure and giving different cultures and historical periods their due. The results had a peculiarly euphonious quality: Mangala Vallis (Sanskrit), Kasei Vallis (Japanese), Ma'adim (Hebrew), Mawrth (Welsh), Nirgal (Babylonian), Dzigai (Navajo), Tiu (Old English), and Simud (Sumerian). Eventually this store ran dry, just as the store of astronomers available for crater-naming purposes had, and valleys began to be named for earthly rivers, both modern

(Loire) and classical (Rubicon), the only proviso being that no name that Lowell had used for a canal could be reused for a valley. (Oddly, though, canal names could be used for the canyons classed as chasmae. Go figure.)

Armed with this knowledge, let us head out into space and park ourselves ten thousand miles or so above the Martian north pole. The most basic fact of Martian geography is that while the southern hemisphere is dominated by cratered highlands, much of the northern hemisphere is covered by smoother, lower-lying plains. These northern lowlands are laid out below us in a roughly equilateral triangle, centered on the pole. The triangle's points mark the places where the lowlands reach furthest to the south; its sides show where higher ground reaches furthest north. Within this triangle, the surface is two miles or more below the notional "zero" altitude that serves as a surrogate sea level on this sealess planet. The texture of the plains changes from place to place: In some places they are phenomenally smooth, in others hummocky; some areas are cut into polygons by cracks like those in dried mud, but far, far larger; some of the surface is covered in dark dunes of volcanic sand, some is strangely mottled.

Imagine a circle drawn inside this triangle so as to touch all three sides. This circle represents the planet's fifty-fifth parallel, more or less. Within the circle sits the region called Vastitas Borealis,* a dark lowland area that completely encircles the planet; at its center is the north polar ice cap, its ice arranged in strange swirls like a frozen hurricane, the whole thing raised above the plains on a plinth formed over billions of years by the dust trapped in the ice that freezes there. The plinth's edges reveal a bizarre lamination, a layered terrain with no earthly analogue. A striking canyon, Chasma Boreale, cuts from the ice cap's edge toward its center in a sweeping curve.

The three points of the northern triangle—the places where the plains reach closest to the equator—are Amazonis Planitia (about

*The reason I didn't mention the term "Vastitas" in my run through the IAU lexicon is that it's a one-off oddity: Though there are sixty-seven assorted planitias and planums on the solar system's other planets and moons, there is not a single other vastitas anywhere on Mars or off it.

160° W, connected to Vastitas Borealis by an intermediate plain called Arcadia Planitia), Chryse Planitia (about 30° W, connected to Vastitas Borealis by Acidalia Planitia) and Isidis Planitia (about 90° E, connected to Vastitas Borealis by Utopia Planitia). Between Chryse and Isidis sits a heavily cratered region called Arabia Terra. Between Isidis and Amazonis sits a volcanic province called Elysium. And between Amazonis and Chryse sits the far bulkier volcanic province of Tharsis.

At this point, let's move a quarter of the way around the planet from our perch over the pole to look down on Tharsis from somewhere high above the equator. The lack of mountain ranges, coasts, and distinctive climates or types of vegetation, not to mention the complete absence of politics, leaves no clear way of defining regions on Mars, and so it's hard to say with any precision where Tharsis begins or ends. The IAU naming committee had originally intended to name regions that corresponded to features seen on the earthly albedo maps, but when it found that the light and dark marking seen from Earth reflected the distribution of wind-borne dust more than anything more substantial it gave up. For managerial purposes, the surface of Mars is normally subdivided along the arbitrary lines used by the USGS to define its thirty mapping quadrangles (another echo of the American West). Quadrangle 9 is called the Tharsis quadrangle, but its 1,700,000 square miles cover about a quarter of what can be thought of as "Greater Tharsis," the continent-sized topographic rise between Amazonis and Chryse. This huge volcanic bulge, completely dominating the planet's western hemisphere, is the second most basic fact about Martian geography after the division between northern lowlands and southern highlands.

Three volcanoes form a line close to the crest of the bulge, moving from southwest to northeast. When they saw their summits sticking out of the global dust storm the *Mariner 9* team called them South Spot, Middle Spot, and North Spot; now they are Arsia Mons, Pavonis Mons, and Ascraeus Mons, with Arsia at the southern end of the line, Ascraeus at the northern, and Pavonis right in the middle and slap on the equator. From our imaginary viewpoint we are staring straight down into its crater. The mountains are spaced about five hundred miles apart, each is at least two hundred miles across, and they stand between six and nine miles proud of the surrounding plateau, which is itself three to four miles above the notional "zero" altitude. Massive lava flows cover the plains at their feet. What appear to be vast landslides extend off their flanks toward the northwest, where the Tharsis plateau starts to slope away. Farther in this direction, just off the edge of the bulge, sits Olympus Mons, grander still. Its broad peak towers more than fifteen miles over the low-lying plains of Amazonis that spread out to the west.

Due north of Tharsis Montes, and about twice as far away as Olympus, the northern end of the rise is dominated by a volcano lower in profile than the others but much, much wider. The lava flows of Alba Patera cover more ground than those of the other Tharsis volcanoes combined: more than one million square miles, almost the surface area of India. In terms of volume, as opposed to height, Alba is probably the largest volcano in the solar system. Not that you'd know it from the surface: The average slope of its sides is about 1 percent. Scattered between Alba, Olympus, and the three spots are various smaller volcanoes. There are also impact craters, but not very many of them. The surface is bright. Mars has no soil as such, because soil, to a geologist, is in part organic; instead it has "regolith," stone broken into sand and dust and gravel by billions of years of asteroid impact (a process bizarrely known as "gardening"). Tharsis is bright because its regolith and bedrock are covered with fine dust. In places the surface is marked by strange wandering paths left by dust devils, mini-tornadoes that pump the fine dust into the sky.

It is this suspended dust that makes the Martian sky yellow brown, rather than an earthly blue—or, as is often claimed, pink. The idea that the Martian sky might be pink came from early data from the *Viking* landers. The *Viking*s had color cameras that could send back red, blue, and green versions of every scene; there were little reference charts attached to the landers so that the balance between the different colors could be properly established back on Earth. Unfortunately these charts were not visible in the first images from *Viking 1*, and so the color balance had to be done by guesswork. The result was a brownish-red surface with a grayish sky that, in some prints, came out as blue, and this was how Mars looked in the first pictures released to the press. A couple of days later pictures of the color chart on *Viking 1* were obtained, and James Pollack—who had a keen interest in how much dust there was in the atmosphere—recalibrated the camera. Now the soil and rocks were red and the sky a dusty pink. This new unearthliness was not universally popular. According to William Poundstone's excellent biography of Carl Sagan some people actually booed Pollack's announcement, and the son of one member of the *Viking* biology team went round JPL retuning video monitors to make the sky look blue again. The shock wore off, though, and the pink skies soon became entrenched in the popular picture of Mars.

However, more painstaking calibrations that took into account all the oddities of the camera's color filters showed that there was too much red in the pink-sky pictures. Using the lighting and color definitions set down by the *Commission Internationale de l'Eclairage* and the National Bureau of Standards–Inter-Society Color Council, the *Viking* lander team gave its final verdict: The soil at the two landing sites was "moderate yellowish brown" and the sky "light to moderate yellowish brown." Twenty years later the *Mars Pathfinder* camera team found exactly the same colors in its most considered assessment of the issue, adding that the rocks, which the *Viking* team had been unable to treat separately, were a "dark grayish yellowish brown." Still, as Peter Smith of the *Pathfinder* team points out, it is hard to stop people from thinking of the rocks as red and the sky as pink: "There's a tremendous bias to make Mars red." Indeed Ray Arvidson, who led the *Viking* lander camera team and took part in its final color calibrations, still remembers the sky as salmon pink: in conversation.

To the east of Pavonis, where the Tharsis bulge is at its highest, there is a strange lacy network of relatively deep canyons. The *Mariner 9* team called it the Chandelier: Because it was close to what Schiaparelli had called Noctis Lacus, the IAU called it Noctis Labyrinthus, the Labyrinth of Night; they're both good names. To see the Chandelier, you have to imagine it hanging north from the southern hemisphere, then its interconnected canyons become hanging crystals. To see the Labyrinth you just have to imagine being inside one of those canyons, trying to escape. At its eastern end the maze opens up into the Valles Marineris.* The first stretch of Marineris is made up of two parallel canyons, Ius Chasma and Tithonium Chasma, gouged into the crust as though by God's fingernails. Tithonium peters out, but after six hundred miles or so Ius opens up into a set of shorter, wider canyons, all following the same rough east-west line, all interconnecting: Melas Chasma in the south, continuous with Ius; Candor in the center following the line of Tithonium; Ophir in the north. Within the canyons are strange tablelike mountains, some of which rise almost to the height of the surrounding plains. The canyon walls rise up three to four miles from their floors and, even though they are not sheer, from below they must look grander in their way than anything on Earth. Given the width of the canyons, though, from the foot of one great wall you might well not see the other; if you did, it would just be as a doubling of the horizon. Landslides that could cover counties spill down from the walls and across the floors.

The southern part of Greater Tharsis is in comparison rather dull; a set of four high plains, Daedalia Planum in the west below Ascraeus Mons, Syria Planum, and Sinai Planum clustered under Noctis Labyrinthus, one to each side of the chain from which the Chandelier hangs, and Solis Planum to the south of them.† Between Daedalia and the other three plains is a strange, raised, folded, and faulted terrain called Claritas Fossae, which runs roughly north-south; its faults are part of a great family of faults throughout the region. To the north these faults are most obvious where they run across the lower flanks of Alba Patera, seeming to part around the volcano's central region like the grain of a piece of wood running

*The second and last major exception to the naming conventions: Valles Marineris is not a set of eroded valleys like Mangala Vallis, Nirgal Vallis and their kin; it is a set of far larger, steep-sided canyons, each individually named as a chasma. Perhaps to underline its exceptional status, it is talked of in the singular despite being clearly plural.

†How, you may wonder, given Schiaparelli's naming conventions, did Syria and Sinai end up so far from Arabia? The answer is that the Syria in question is not the Levantine country, but the island in the *Odyssey* "where are the turning places of the sun" (hence its position near Daedalia and Solis Planum), and that Sinai was named in error when the IAU cleaned up the planet's nomenclature; it was close to a feature Schiaparelli had called Mare Erythraeum, which the IAU translated literally as "the Red Sea," rather than correctly as the Indian Ocean.

past a knot. In the eastern parts of Solis Planum, and in the corresponding plains to the north of Valles Marineris, the smooth lava flows that make up the surface are wrinkled into a set of ridges: This is the terrain that the *Mariner 9* team called the Elephant's Ass. These ridges and faults clearly express the fabric of the land. For the most part the faults run out of the center of Tharsis (as does Valles Marineris) like rays from a stylized sun; the ridges ring the great bulge in concentric arcs.

Time to move another 90° around the planet, this time following the equator to the east. Now we look down on a troubled land called Margaritifer Terra, Schiaparelli's pearl coast. The surface appears to have slumped and shattered. The chaotic regions start at the end of the last of the canyons of Valles Marineris, Coprates Chasma, one of the few features to be named after one of Lowell's canals. Vast channels swing north from this terrain and curve downhill toward Chryse and the northern plains: Ares Vallis, Simud Vallis, Tiu Vallis. Farther west more such channels sweep down from chasms and chaoses to the north of Valles Marineris, cutting through the cratered terrain called Xanthe Terra as they rush toward the lowlands. West of Xanthe a smoother plateau, Lunae Planum, sits undivided, its surface wrinkled by Tharsis-centered folds. The western edge of this plateau, two-thirds of the way back to our previous vantage point above Pavonis Mons, drops away sharply in a spectacular feature called Echus Chasma, a cliff six hundred miles long and three miles high in some places, apparently far steeper than the sides of the Marineris canyons. If worlds had walls, this is what they would look like.

East and south of the chaos at the end of Valles Marineris, the terrain becomes more typical of the southern highlands, the most widespread style of landscape on the planet. It is heavily cratered, with some craters much larger than any on the plains of Tharsis or the northern lowlands. Directly to the east of Margaritifer, the landscape is somewhat subdued, the crater rims and the plains between them smoothed: This is Terra Meridiani, the southern part of Arabia Terra, a large but relatively low-lying shelf of the southern highlands that extends up into the north between 10° W and 80° E. On the equator, slap in the middle of Arabia, sits the largest of the craters named for astronomers: Schiaparelli, 290 miles across, a hole half the size of Italy.

To the south of Terra Meridiani things are sharper and clearer. This is Noachis Terra and it shows the southern highlands in their archetypal form, its craters sharper, the space between them fractured and faulted. South and west of Noachis lies the basin called Argyre Planitia, almost twice as wide as Schiaparelli. A basin, in astrogeology, is the result of an impact that creates shock waves capable of doing more than just opening up a circular hole, shock waves that can bend and warp the planet's crust in complex ways. Concentric arcs of uplifted crust ring Argyre like frozen ripples. The crust below the bull's-eye basin is thinned: The surface lies lower than it does in the surrounding area and the base of the crust—the depth at which the rocks of the crust give way to the chemically distinct rocks of the mantle that lie beneath them—probably rises higher here than it does elsewhere. Argyre's eastern rim is interrupted by a smaller crater 130 miles across, Galle—a potshot that just missed the bull's-eye. Within Galle, secondary features have conspired to make an almost perfect smiley face, with two oval eyes and a grinning arc below them. (This is not the "Face on Mars" that some see as a monument left by some vanished race; that is a much smaller affair on the edge of Chryse.)

Moving another 90° east along the equator, we pass from one edge of Arabia Terra to the other. Below us the northern plains have again reached almost to the equator in the form of Isidis Planitia, an impact basin largely enclosed by the highlands but that, in its northeast quadrant, opens into the far more extensive lowlands of Utopia Planitia. West of Isidis is a comparatively smooth volcanic plain, Syrtis Major Planum, its surface dark and scoured of dust: the Hourglass Sea that was the first feature on the face of Mars to become a landmark to astronomers. To the south and east lies a volcanic plain of a similar texture, Hesperia Planum, centered on Tyrrhena Patera, a sprawling volcano fittingly known as the Dandelion in *Mariner 9* days. With these exceptions, the rest of this part of the southern hemisphere looks much like the rest of the highlands—until, about 30° south of the equator, your eye lights upon Hellas Planitia.

Hellas is almost three times as wide as Argyre, to the west of it, and its floor is more than five miles lower than the level of the surrounding cratered plains. On MOLA's maps of Mars, color coded by altitude, Hellas sits in the highlands like an amethyst half the size of Australia set in a ring of ruddy gold; the surface within the basin is lower than anywhere else on the planet. It has been suggested that impacts this large could affect the surface half a planet away: Shock waves racing out through the planet would be focused on the point antipodal to the impact, rupturing the crust there. This seems to have happened on Mercury, where the crust opposite the Caloris Basin has been distorted in various strange ways. The shock waves from Hellas, almost twice the size of Caloris, would have met halfway around the planet on the northern edge of Tharsis—at the site of Alba Patera, that vast, low, and bizarre volcanic structure. Whether this is coincidence or not, no one can say.

North of Isidis, in the regions called Nilosyrtis Mensae and Protonilus Mensae, is located what the *Mariner 9* team called the fretted terrain, where the highlands seem almost etched away by the northern plains. To the east of Isidis, Utopia eventually gives way to the modest volcanic rise centered on Elysium Mons, similar in size to the big three atop Tharsis. Flanking it sit Hecates Tholus and Albor Tholus, mountains of witchcraft and whiteness.

A last 90° jump takes us due south. The southern ice cap sits on a high, off-center plain, Planum Australe, which partially covers a large impact basin and is ringed by what look at first like typical highland landscapes. However, the craters in the highland materials have a peculiar relaxed appearance, slumped and smoothed, which is widely thought to be due to the presence of ice in the ground; the few craters at high latitudes in the northern plains show similar effects. Though the presence of widespread ground ice throughout the high latitudes was not accepted until the 1980s, it is hard now to look at the planet without seeing its traces. The fretting of the northern edge of Arabia seems due to the action of ice in the rock: So does similar fretting within the southern highlands south of Hellas, and the strange polygonal fractures seen in some of the northern plains.

Amid these icy features is a strange set of straight lines at right angles that the *Mariner 9* team referred to as "The Inca City," something that got one of their number, Bruce Murray's student Jim

Cutts, into a little trouble. Though now the most naturalized of Californians, Cutts was born and raised in North Wales and when he got to Caltech various aspects of American life were mysterious to him. Mert Davies remembers him deeply perplexed at the mixed drink orders he received when pressed into service as a barman at one of Murray's parties; perplexed, but eager to please. When Cutts was called up by a journalist asking whether the Inca City might, in fact, really be a city, he explained diligently that it was out of the question. When pressed again and again on the possibility that, even if this was not a city, there might be cities on Mars, perhaps deep underground where no space probe could see, he reluctantly admitted that, while there was no evidence to support such a claim and it would be wildly surprising if it were the case, it could not be utterly and completely ruled out. The journalist thanked him and rang off, and Cutts thought little more about it. He felt he'd got the point across, that the journalist had seemed to understand and that the paper sounded reputable. Cutts had come across the *Philadelphia Inquirer,* and so had to assume that a *National Enquirer* would be much the same, possibly more prestigious. He did not learn the error of his ways until, a week or so later, he was awakened by a call from Joshua Lederberg, a Nobel Prize–winning biologist who had taken a great interest in NASA's extraterrestrial biology research—it was he who dubbed the endeavor "exobiology"—and who was keen to make sure that this research was never tainted by sensationalism. Lederberg was not amused. When he got to a supermarket checkout and saw what had been made of his careful words, neither was Cutts.

Planum Australe itself, standing proud of the deformed terrain around it, is made up of layered deposits similar to those seen at the north pole. These southern deposits stretch farther than their counterparts to the north, covering three times as great an area and reaching up as far as the seventieth parallel in one direction; it was on this outlying part of the south polar deposits that *Mars Polar Lander* was meant to land. However, while these deposits cover a much greater area than the north polar deposits, the actual polar ice cap is smaller. The difference is due to Mars's orbit, which is much more eccentric than that of the Earth. During the southern summer, when the south pole is in almost continuous sunlight and the north pole in shadow, the planet is considerably closer to the sun than it is in southern winter/northern summer and this has its effects on the cap. Both caps have a seasonal component—thick carbon dioxide frost frozen out of the atmosphere in winter—and a permanent component. Because the southern summers are relatively intense, once the seasonal carbon-dioxide cap sublimes (transforms from a solid to a gas without ever becoming liquid) the permanent cap that remains is quite small. The north pole's permanent cap, never heated as much, is far larger than its southern partner, and is composed of frozen water, not carbon dioxide.

In time, this will change. Mars's axis of rotation, like that of the Earth, is slowly precessing—tracing out a circle around an imaginary line perpendicular to the plane of the planet's orbit. In twenty-five thousand years, it will be the north pole that faces the sun when Mars is at its closest and the northern hemisphere will have hot, short summers like those the south experiences today. This regular pattern of climate change is thought to explain, in part, the peculiar layered terrains around the poles; if they are created from dust left behind when the seasonal cap sublimes away, their layers correspond to changes in the size and behavior of the ice cover over time. The precession effects would drive some of these changes; so would changes in the shape of the planet's orbit, which take place in a complex set of cycles between a hundred thousand and two million years long. Perhaps most important are changes in obliquity—the angle between the planet's axis of rotation and a line perpendicular to its orbital plane. Currently Mars has an obliquity similar to that of the Earth, 25.2° as opposed to the Earth's 23.5°. But the Earth's obliquity stays fairly constant, while Martian obliquity varies greatly, shifting chaotically over the course of the planet's history from a sitting-straight-up 0° to a spectacularly slouched 60°. On the Earth, that would give London seasons that cycled between icy winters of darkness at noon and tropical summers with midnight suns, while the equator spent most of the year in temperate coolness.

The Earth is saved from such drastic rocking back and forth by its moon, an exceptionally large satellite that steadies the planet's axis of rotation. Mars has no similar stabilizer; its two irregularly shaped moons are far too small to exert a steadying gravitational influence.

Phobos, the inner moon, is just seventeen miles long and orbits only 3,700 miles above the planet's surface, completing its circuits in just seven and a half hours. With an orbit so low and almost exactly in the plane of the equator, it is never seen from the polar regions and spends quite a lot of its time eclipsed in the planet's shadow. Deimos is smaller still, less than half the size of Phobos; it orbits farther out and at a more sedate pace, coming full circle every thirty hours. The fact that Deimos moves around Mars only slightly slower than Mars spins on its axis makes its passage from east to west remarkably leisurely; it lingers in the sky for sixty hours at a time. Phobos, by way of contrast, races around its orbit considerably faster than the planet spins on its axis, which means it rises in the west and sets in the east just four hours or so later. Ray Bradbury was unaware of this oddity of the Martian sky when writing *The Martian Chronicles,* as he recalled during the "Mars and the Mind of Man" colloquium at Caltech in 1971:

> A few years back, one dreadful boy ran up to me and said, "Mr. Bradbury?"
> "Yes?" I said.
> "That book of yours, *The Martian Chronicles*?" he said.
> "Yes," I said.
> "On page 92, where you have the moons of Mars rising in the East?"
> "Yeah," I said.
> "Nah," he said.
> So I hit him. I'm damned if I'll be bullied by bright children.

This apparently contrary behavior on Phobos's part will eventually cost it dear. Tidal forces normally weaken the bonds between planets and their moons over time; the Earth's moon is receding at about an inch and a half a year, gently slowing the planet's rotation as it does so. However, because its orbit outstrips Mars's own rotation, these forces have the opposite effect on Phobos, dragging it ever nearer to the planet below. At sometime in the next hundred million years it will be torn apart as it gets too close, some of it crashing to the surface, some of it creating a ring of debris in orbit.

A thin Saturn-like ring might liven the Martian sky up a little; at the moment it lacks spectacle. Not only are both the moons small;

they are also made from extremely dark material. From the surface Phobos appears a quarter of the size of the full moon as seen from Earth, despite being sixty times closer; it can offer only 10 percent of an earthly full moon's light. That said, it is still a bright object—three hundred times brighter than Venus as seen from Earth, and thus clearly visible by day or night.

So there you have it: Mars, a planet with mountains larger than its moons. It is a small place by earthly standards, its equatorial girth only a little more than half the Earth's, its mass ten times less. Yet its features dwarf those on this planet in almost every way. From the depths of the Pacific to the heights of Everest is just over eleven miles; the heights of Olympus rise almost eighteen and a half miles above the depths of Hellas. The mountains of Mars are the size of American states, its grandest canyons are continental, and its cliffs would challenge many aircraft. Plains the size of Canada lie flooded by undivided sheets of lava. Great basins mark the impacts of asteroids far larger than the largest recorded as hitting the earth—the creation of Hellas would have boiled half the Earth's oceans dry.* Yet in all the ways that Martian features dwarf the Earth's, the most striking is not to do with their size—at least not directly. It is to do with their age.

Mars is a dry, dusty desert; as such we expect it to be old. Deserts are antique lands in which we expect the ancient to be preserved, places where we find rose-red cities half as old as time, vast and trunkless legs of stone and other relics of the most distant past. This expectation of age has colored our images of Mars since Lowell. H. G. Wells's Martians were vastly intelligent because they had evolved for longer than we had; the dried, dead seabeds of Edgar Rice Burroughs had seen whole civilizations fall. These ancient civ-

*Some have transferred the giganticism of Martian landscapes onto the planet's inhabitants; Lowell imagined that weaker gravity would allow Martians to be three times taller and nine times stronger than humans. In Kim Stanley Robinson's novel *Red Mars* the colonists invent a mythical "Big Man" whose name gets attached to outsized features: "Big Man's Toothbrush," "Big Man's Shaving Bowl." Applying the nineteenth-century biologist Karl George Bergmann's observation that animals in related species tend to be bigger in colder climates, Carl Sagan used to speculate that Martians might be huge, polar bear–like things.

ilizations, often seen as past their prime, played into a strand of planetary orientalism that imagined a Mars of despots and decadence. (Lowell himself, in one of the popular books he wrote on Japan and the Far East before turning his attention to the planets, compared oriental civilization with "what we see cosmically in the case of the moon, a world that died of old age.") In many fictions the Martians are all already gone; the only links human explorers have to them are archaeological. The lost civilization imagined around the pharaonic "Face on Mars" seen by some in *Viking*'s pictures of the Cydonia tablelands east of Chryse fits into this aesthetic perfectly, which must explain part of its attraction: Mars evokes ruins because it is in some way a ruin itself.*

But such expectations do no justice at all to the true ages involved. Archaeologists deal with years in their thousands or tens of thousands; most geologists on Earth handle years in their tens and hundreds of millions. For Martian antiquity, though, we must graduate to the billions. Mars, the Earth and the rest of the solar system were formed an astonishing four and a half billion years ago. In terms of space, the solar system and its planets are quite small—a human being can walk around the Earth in a decade; humanity was launching spacecraft out of the solar system within two decades of first putting them into orbit. But in terms of time they are immense. They occupy far less than a billionth of one galaxy among billions, but they stretch a third of the way back to the Big Bang itself. And while the Earth hides this extraordinary age, Mars wears most of the all-but-countless years quite openly.

On the Earth, erosion and plate tectonics endlessly destroy and re-create the planet's crust. Mountains rise and fall like waves; ocean basins open and close; continents are smashed together and worn down. In its youth, more than four billion years ago, Mars may have seen similar flux. But Mars is small, and will have lost the internal heat that drives such activity quite early on; since then it has slowed

*If Mars was always the old planet, cloud-covered Venus frequently used to be seen in science fiction and speculative science as a juvenile Earth, a swamp of innocence—and possibly dinosaurs—corresponding to the Earth's Carboniferous or Cretaceous, with civilizations, if any, at the cool poles.

down immensely. If the dinosaurs had had a space program, back when the rocks of Everest lay undisturbed on the floor of an ocean not yet Indian, they would have discovered an Olympus Mons pretty much as big as the one that peeped out of the dust clouds at *Mariner 9.* The volcano was not much shorter than it is today when those dinosaurs' ancestors first crawled onto the land. It may have been an imposing presence a billion years ago. Less than 10 percent of the Earth's surface is that old; yet Olympus Mons is thought to be a relatively new bit of Mars.

Slow growth is matched by even slower erosion. Everest will fall and some new mountain will claim its laurels in the next great range that rises. This has happened hundreds of times in the history of the Earth, and some of the mountains thrown up may have been significantly grander than today's peaks (though never as grand as those of Mars, the Earth's higher gravity and thinner crust make such enormities impossible). On Mars, though, the things we see today are the grandest things the planet has ever had to offer. For billions of years Mars's mountains have not been ground down; for billions of years, Mars's basins have not been filled in and the edges of its canyons not leveled off. The most prodigious exuberances in the aeons since its turbulent youth are still on display today. As is the way in deserts, nothing is forgotten.

The immensity of Martian landforms is not a symptom of some planet-wide creative potency—quite the reverse. Over the past four billion years the Earth's volcanoes have spewed forth hundreds of times more lava than their Martian counterparts. The Earth's crust is continually being torn apart along rifts far longer than Valles Marineris—rifts that eventually widen into oceans. The creation of the great channels that scour the surface of Mars involved the movement of a hundred trillion tons of rock; the Earth's rivers deliver more sediment to its oceans every millennium. For all its relatively low relief, the Earth is in perpetual ferment; it takes real effort for a planet to grind itself down.

Mars cannot make that effort; nothing much happens there. As the poet Frederick Turner puts it in *Genesis,* an epic poem about the transformation of Mars by humans:

Time here is cheap. A billion years can pass
Almost without a marker; if you bought
A century of Marstime in the scrip
And currency of Earth, you'd pay an hour
Or half an hour of cashable event.

Even at so poor a rate of exchange, though, things add up if you wait long enough. Mars has centuries to spare, and all that happens leaves its mark. On Earth, geological time is deep and mysterious, its strata for the most part buried beyond inspection; on Mars vast tracts of time sit on the surface, exposed but little altered. There are landscapes on Mars that have not changed materially in a billion years. Learning the features on the face of Mars may take a little effort, but it's knowledge with no use-by date. Most of what I've just described will outlast the continents that grace the Earth today. Much will still be there when the sun burns out.

Maps and Multiple Hypotheses

> A map seems the type of a conceptual object, yet the interesting thing is the grotesquely token foot it keeps in the world of the physical... being a picture of the degree that that sacrament is a meal. For a feeling of thorough transcendence such unobvious relations between the model and the representation seem essential, and the flimsy connection between acres of soil and their image on the map makes reading one an erudite act.
>
> —Robert Harbison, *Eccentric Spaces*

"They don't know how goddamn wrong they can be," Dave Scott told me over a stiff glass of bourbon as Flagstaff fell into the Earth's shadow one evening. Scott is, in Lowell's phrase, a man of more worlds than one. After getting his first degree from Caltech in the late 1930s, he spent the next thirty years in the oil business, mapping out fields all over the world. Sometimes he made his employers money. Sometimes, though, he didn't; like all oil geologists he'd sometimes persuade the company to buy a lease and drill a hole and have nothing whatsoever to show for it—except an appreciation for how goddamn wrong he had been.

In the 1960s he changed tack. He went back to school, got a graduate degree, and joined the Survey's Astrogeology branch. In the mid-1970s he found himself in charge of the geological mapping of Mars. He watched as experts from the Survey and beyond used the techniques that Shoemaker had pioneered on the moon to try to make sense of the thirty quadrangles into which Mars had been chopped. And he realized how different his new world was from his old one. On Mars it was possible to get an idea and pursue it unencumbered by any risk of the certainties that follow from drilling a dry hole. The planetary geologists could have ideas as wild as they liked, ideas that flatly contradicted other people's views of the same issues, and there was no sure way of telling the good from the bad. They didn't know how goddamn wrong they could be. At times this frustrated Scott, but he could appreciate the liberation involved.

In the late 1970s, Scott and a British-born colleague named Mike Carr took the thirty studies of Martian quadrangles and synthesized them into a geological map of the whole planet. When all the data were in from *Viking*, Scott, Ron Greeley of Arizona State University, John Guest of University College London, and Kenneth Tanaka, who had joined Scott in Flagstaff, edited and expanded the original geological map into a three-sheet 1:15,000,000 synopsis of the planet that was also available in digital form. This is now the basic document of Martian geology. It's not definitive, it doesn't show everything, and in places it's doubtless wrong. But it's where people start from.

The 1:15,000,000 geological map, like most geological maps, looks like a madman's quilt. It recognizes ninety-four different rock units. Some are widespread: One type of ancient cratered terrain is shown covering more than a tenth of the planet's surface, twice the area of the United States. Other units are no more than a tenth the size of Alabama. Some units crop up in widely separated places: Lunae Planum, just east of Tharsis, is shown as being the same sort of "ridged plain" as Hesperia Planum, halfway around the planet. Each unit has its own distinctive color, arranged in loose families of similar tone or hue. The ancient cratered units of the southern highlands are bronze, brown, and tan; volcanic plains are light violet and dusty pink, crater floors a pus yellow. The lava fields of Tharsis are arranged in tones of pink and heather and magenta, with Alba ringed in a deep steely purple. The volcanoes of Elysium are red as wine, evoking the fires within; the floors of the great channels south of Chryse are a range of blues in honor of the waters that seem to have carved them; the plains they lead to are broad expanses of green and turquoise, colors echoed in a dustier palette to represent the rocks that line the Hellas basin. At the north and the south, the polar layered terrains stand out in a startling azure.

Defining the units was one thing; saying what they represented was quite another. Geologists—especially older geologists—like to say that the best geologist is the one who's seen the most rocks, and there's a lot of truth to that. Total number of Martian rocks seen by Martian geologists in the 1970s? None. By 1976 there were television pictures of surface rocks from *Viking,* and there were interesting data in those images. But for geologists, seeing a rock is much more than looking at a single image of it from a single point of view. It's walking around it, chipping away at it, examining it with hand lens and microscope, scratching other rocks with it, trying to find other similar samples, understanding how it relates to other rocks around it. They'll grind them up and taste the grit when they think it might help. Geologists are about the least passive observers you could ever hope to find. Faced with Mars, though, they basically had to make do with images and interpretation.

You might think that sophisticated spacecraft could provide some diagnostic data about the rocks they are looking at, and so they can—up to a point. Spectroscopic analysis of light reflected from the Martian surface can be used to identify the minerals that make up the rocks, and by the mid-1970s it was already becoming clear from the spectrometer on *Mariner 9* that the dark regions of Mars appeared to have the same mineralogical make-up as basalt, the dense volcanic rock that makes up most of the crust under the Earth's oceans. The lighter, redder regions appeared to be similar, but covered in iron-rich dust. In the late 1990s the spectrometer on *Mars Global Surveyor* confirmed the overall similarity to basalt while also finding evidence that over some parts of the planet the surface minerals are more similar to a subtly different type of volcanic rock, called andesite.

Vast volcanoes, lots of basalt: It would be easy to assume that Mars was just a ball of lava. And maybe much of it is. But spectroscopy only measures minerals and minerals aren't everything. If you grind down basalt into sand, and that sand then sets into sandstone, the sandstone may well look the same as its igneous parent to a spectrometer in orbit. But to a geologist in the field the two would look very different, and tell very different stories; a sedimentary unit and an igneous one will have completely different histories. Nor are all basalts the same; they can come out of volcanoes as liquid or as ash, they can stay in localized flows or flood whole basins, they can spread through the crust at depth; they can be eroded in different ways and to different degrees.

Deprived of rock samples, the mappers had to rely on the surface texture to distinguish different units. But even then they could not say for sure what they were looking at. A surface that shows signs of having flowed into place may be a lava field. But it may be some sort of solidified mud or a wind-borne sediment. In some places it's fairly obvious if a rock is sedimentary or igneous. It would seem perverse to insist, for example, that the flanks of Mars's vast volcanoes are covered in flowing sediments, rather than lavas.* In other places, though—on the floors of basins, for example—one cannot be so sure. As a result Mars's geological mappers have for the most part been disciplined into not assuming that they know how the rock unit they are describing was formed, or what it is made of. Their descriptions are kept as free as possible of any interpretative input—preferring what Dave Scott calls the "goofy" terms of pure description, such as "mottled terrain"—lest the maps become dated through allegiance to theories about the nature of the Martian surface later overthrown. This neutrality is part of a tradition that dates back to Gilbert and his contemporaries, a way of separating the facts from the multiple working hypotheses by which they might be explained. Such strictures can be frustrating and younger astrogeologists seem to chafe against them a little. The old guard, which has seen theory after theory come and go, tries to rein them in.

At every border on the maps where one slab of color abuts the next, meticulous eyes have detected a change in surface texture: plains dissected by valleys give way to undissected plains, mottled plains to unmottled ones and so on. Sometimes the boundary is obvious, and sometimes the two units just seem to blend into each other and the boundary becomes a matter of subjective judgment. When maps of two adjacent quadrangles by two different mappers were put next to each other in the early days, many of the bound-

*Just because it's perverse doesn't mean it's impossible. Thomas Gold, an astrophysicist at Cornell who has various iconoclastic views about planets and their interiors, is convinced that Olympus Mons is made of frozen mud.

aries would not line up. "I see more discrepancies than we thought when comparing notes in the [bar of the] Little America last night," one Flagstaff memo from the 1970s laments. Ensuring that such boundaries ended up consistently placed across the various maps was one of Scott's duties as mapping coordinator.

These boundaries are not minor details; they are the heart of the mapping process. Each boundary is a line separating a before from an after, a hiatus in the creation of new rocks that might last a second or a billion years. The point of geological mapping is to tell a story—to turn landscape into history—and the gaps are the story's articulation. It is the way that time sits within the boundaries marked in the rock that makes the geological map such a strange and powerful form of drawing—that gives it a plot and a structure. This representation of time as a function of space is almost unique: but not quite. In his brilliant analysis of what makes comic strips work, *Understanding Comics*, Scott McCloud has this to say about how time is represented between the images in a strip:

> In learning to read comics we all learned to perceive time spatially, for in the world of comics, time and space are one and the same. The problem is there's no conversion chart. The few centimeters that transport us from second to second in one sequence could take us a hundred million years in another. So as readers we're left with only a vague sense that as our eyes are moving through space they're also moving through time—we just don't know by how much.

The insight needs no changes to apply to the reading of geological maps; if we all read such maps as children, they would be as clear to us as the conventions of comic books are.

In most geological situations the boundaries between rock units represent one unit overlying another, just as the frame to the right in a comic strip usually represents events that follow on from those in the frame to the left. In some cases, the relationship was fairly obvious: Where one unit seemed to flow into the other, filling up its depressions, encircling its promontories, then that embaying unit was laid down more recently and overlies buried parts of the earlier unit. Triple boundaries, where three units meet, were much prized, since they reveal the succession definitively and clearly: If one unit overlays the boundary between two others, it must have been laid down later.

These techniques allowed the geologists to define the sequence of events on a local basis, but what of the global history? On the moon, the key to global stratigraphy was ejecta from the large basins, thrown far and wide across the airless surface. If ejecta from the Imbrium basin impact lay over the rock unit you were studying, that unit had to be a pre-Imbrian formation. The moon's geological eras thus take their names from the impacts that mark their beginnings: Nectarian, Imbrian, and so on. On Mars it was not that simple. Though there were big basins their ejecta blankets were not well preserved. Mars does not have much by way of air or weather today, but it has a lot more than the moon has and in the past it seems to have had still more. This means Martian ejecta, especially from the earliest days, is eroded and moved around. Early attempts by some mappers to make the Hellas impact mark a key event in Martian history, to define an Hellasian or Hellanic period, came to naught.

Another technique derived from moon mapping worked well, though: crater statistics. Rather than trying to establish the order in which specific craters and basins were formed, you could simply count the number of craters of a given size in a given area. The more craters there were, the older that patch of surface must be. This crater-counting technique became the main tool by which Martian history was measured, and the planet's geological periods ended up named after the plains where craters were counted, rather than after specific craters and basins. When Scott and Carr came to sum up the geological studies to date in their synoptic map of 1978 they chose an application of this technique that divided Martian history into three parts named not for basins but for cratered plains. The oldest period was the Noachian, named for the heavily cratered region called Noachis Terra that lies between Argyre and Hellas; next came the Hesperian age, named for Hesperia Planum, the large volcanic plain east of Hellas; then came the Amazonian, named for the very lightly cratered plain between Elysium and Tharsis in the northern hemisphere from which Olympus Mons rises.

This system of ages added an historical edge to the basic fact of Martian geography: The northern lowlands were Amazonian and Hesperian in age, and thus younger; the southern highlands were mostly Noachian and thus older. This was not a complete surprise to the geologists: On the Earth and the moon, the only other planetary surfaces studied in detail at the time, there are also fundamental dichotomies of age and altitude. The moon is divided into highlands, which are extremely old, and mare deposits, great floods of basalt that have filled in low-lying impact basins. The maria are often much younger than the basins they sit in; they are much less cratered than the highlands. The Earth is divided into young, low-lying ocean basins with basalt floors and continents made of a granitelike crust that is thicker, lighter, and older (this distinction was what led Gilbert to believe that the Earth had been created from planetesimals of a different density). The idea of a planet with two fundamentally different types of crust was not a novelty.

But that didn't mean it was easily explicable. The processes that had divided the Earth and moon in two were quite different from each other. Which was the better analogy to the split personality of Mars—if either? Over three decades the geologists who study Mars have assembled a variety of working hypotheses to answer this question. And they still don't know how goddamn wrong they may be.

Given that Mars was clearly a battered planet, the idea that the northern lowlands had been created by impacts had an obvious appeal from more or less the moment of their discovery. In the period that followed the formation of the planets four and a half billion years ago there was a great deal of junk left over in the solar system, and the new planets suffered the consequences. The face of the moon dates largely from this time and bears the scars in the form of about forty basins, some of them filled by the flood basalts that make maria.

At some scales most of Mars seemed almost as cratered as the moon, but at the largest scales it didn't. The *Mariner 9* team saw just three obvious large basins: monstrous Hellas and the more reasonably sized Argyre and Isidis. However, telltale signs such as rough circles of uplifted fragments of old crust, concentric patterns of

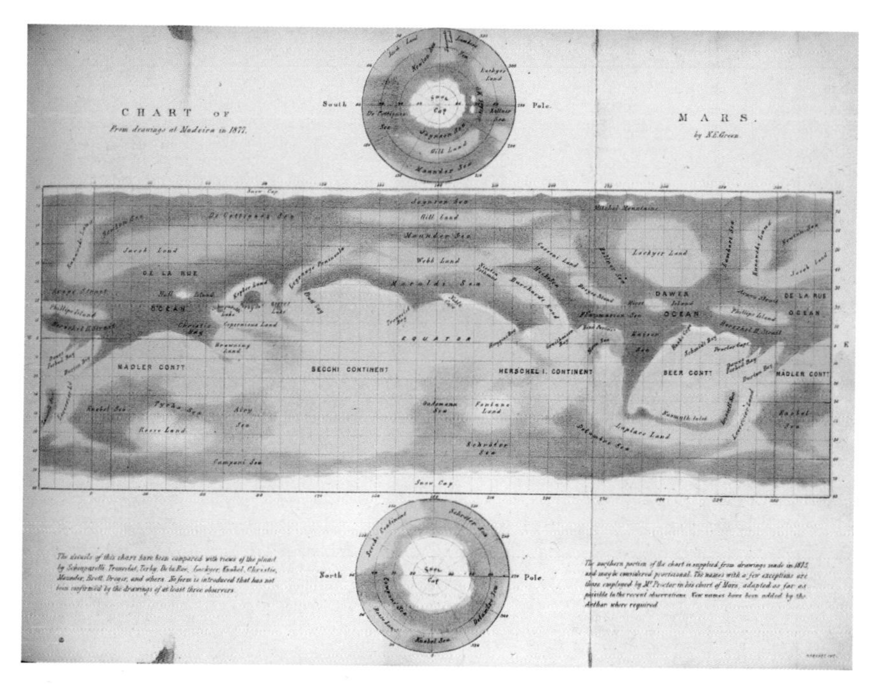

Fig 1: Painted map by Nathaniel Green, from observations in Madeira (1877).

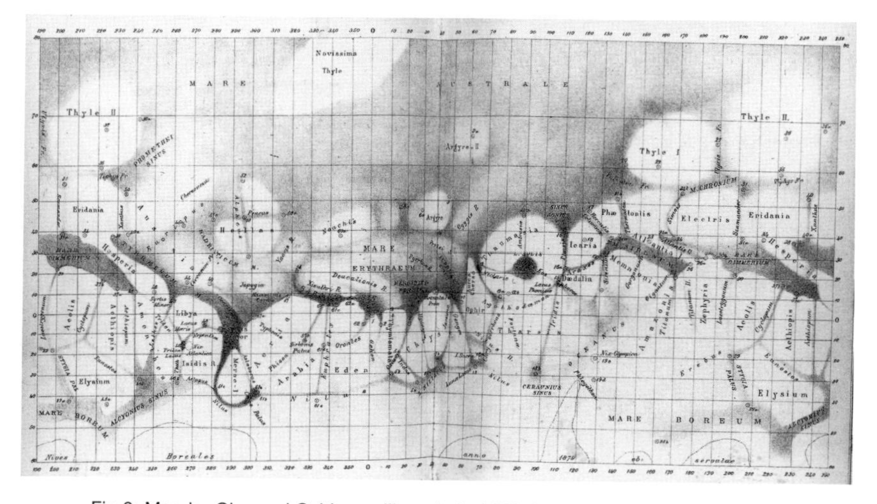

Fig 2: Map by Giovanni Schiaperelli made in 1879. Note the network of “canali.”

Fig 3 (right): Mars wreathed in dust, as seen by the approaching *Mariner 9;* the dark spot is Olympus Mons (1971).

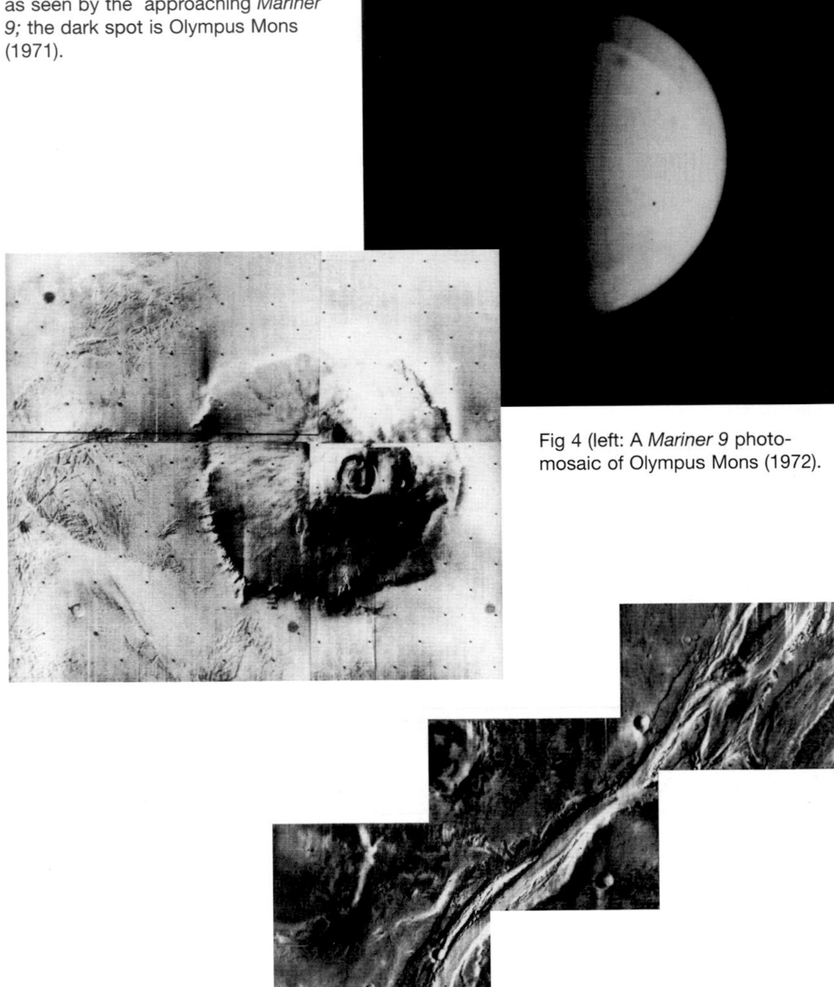

Fig 4 (left: A *Mariner 9* photo-mosaic of Olympus Mons (1972).

Fig 5 (bottom): A *Mariner 9* picture of Mangala Vallis, an outflow channel that drains into Amazonis Planitia (1972).

Fig 6 (top opposite): A *Mariner 9* photomosaic showing the North Pole and surrounding lowlands (1972).

Fig 7 (bottom opposite): An oblique *Viking* view of Argyre Planitia and the southern highlands; the "smiley face" in Galle shows up nicely (1977).

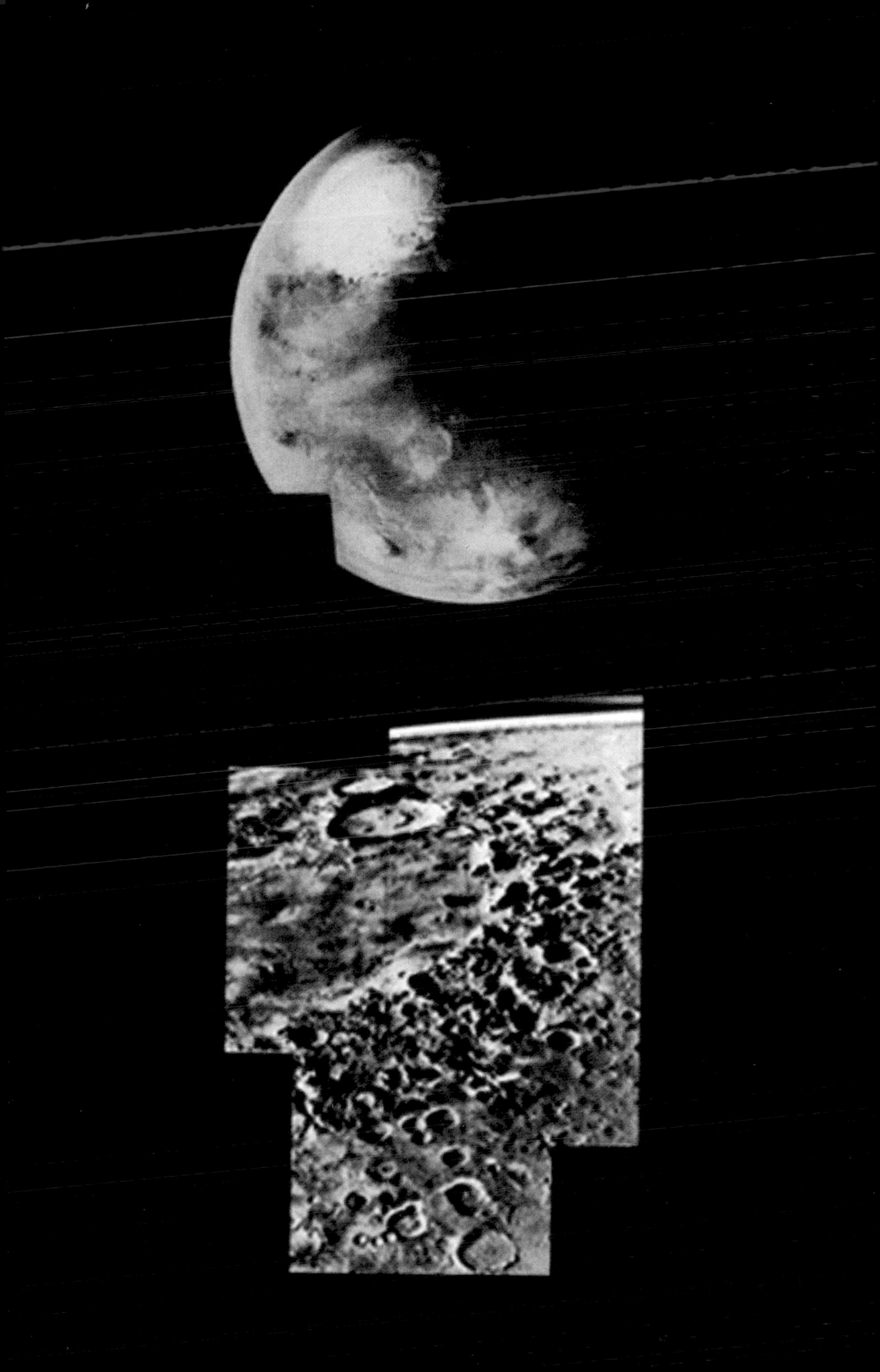

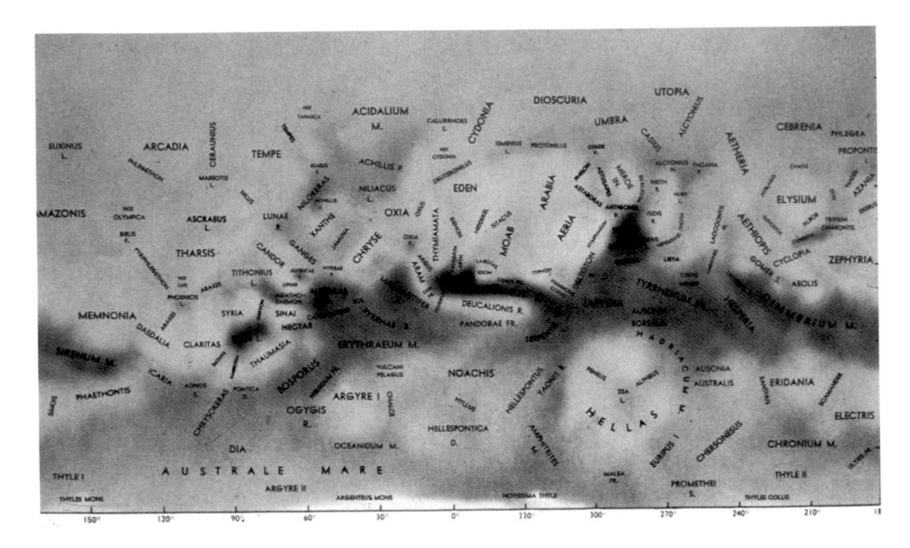

Fig 8: An albedo map prepared by Jay Inge from terrestrial observations (1971).

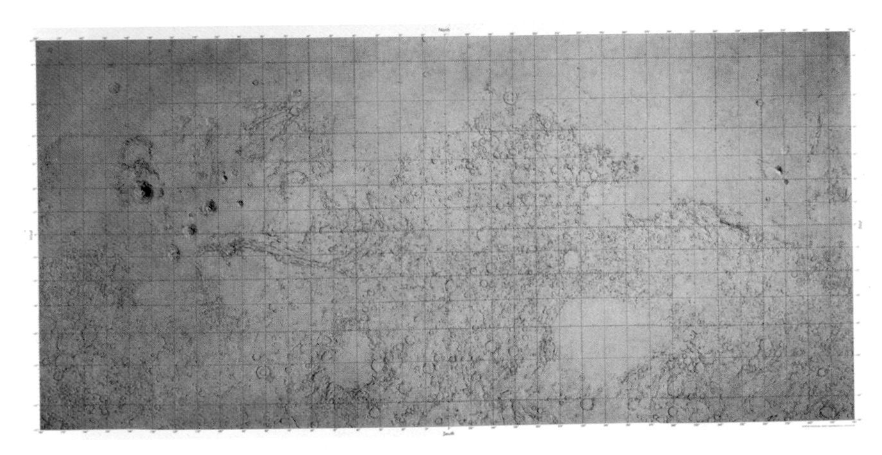

Fig 9: The first shaded-relief topographical map of the whole planet, prepared by Patricia Bridges and Jay Inge at the USGS (1972).

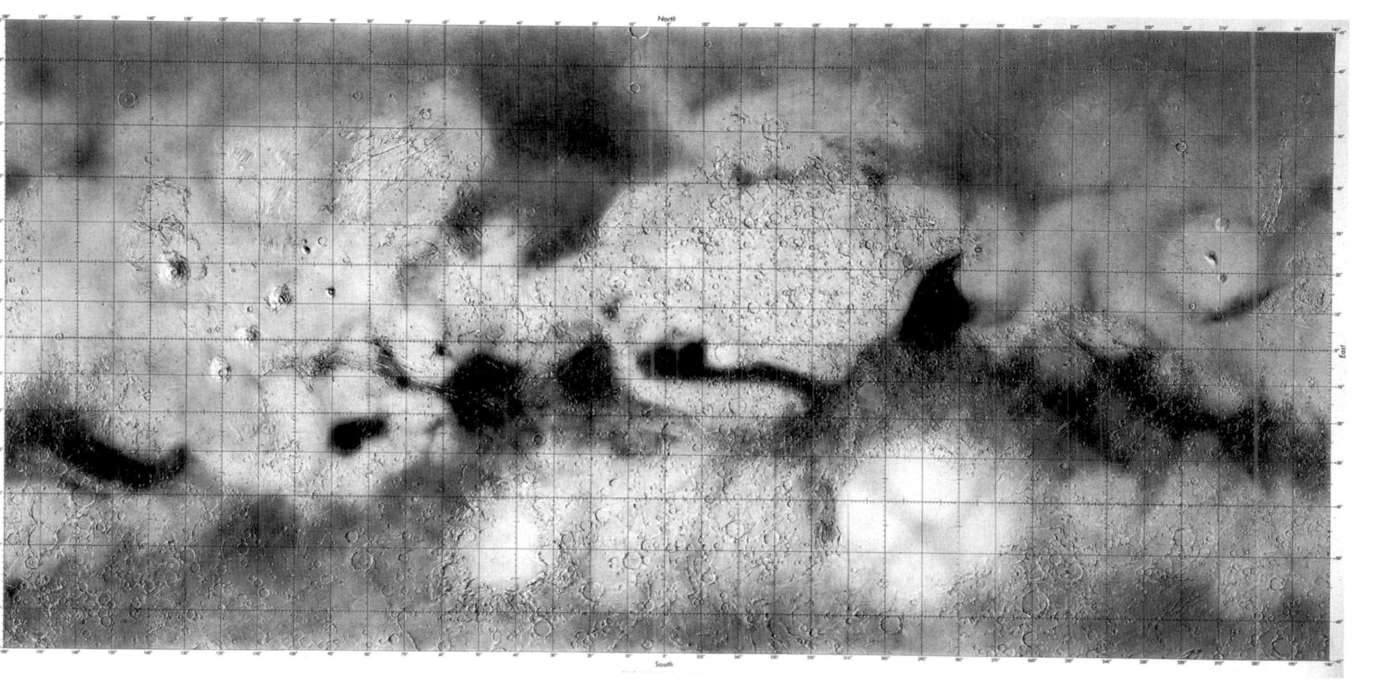

Fig 10: A composite of the previous two maps, prepared by Inge.

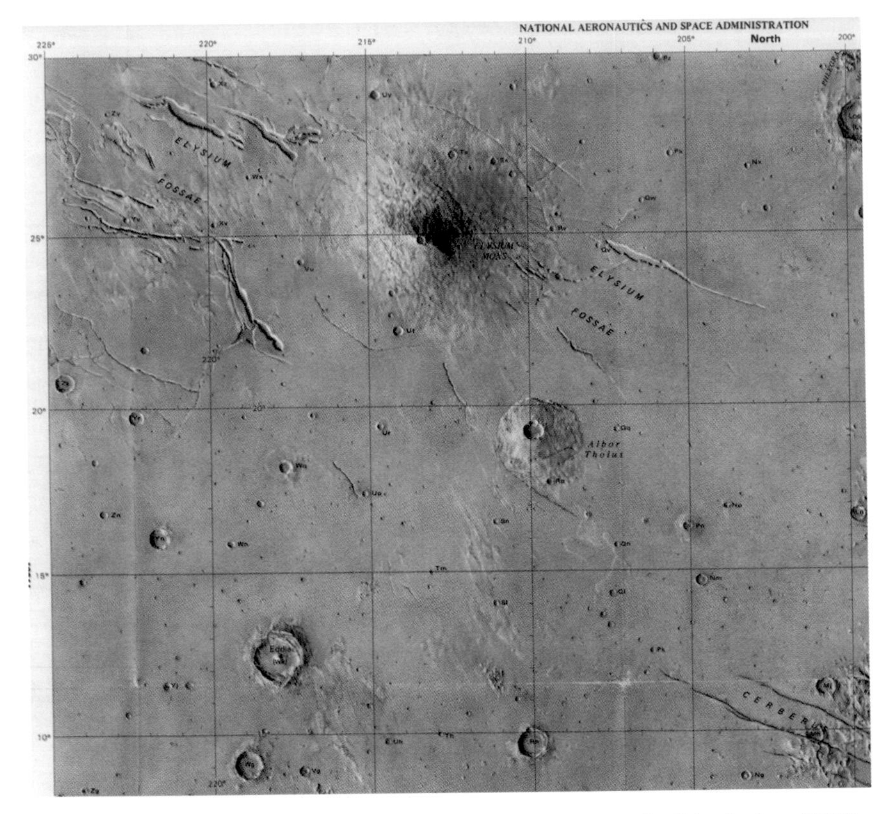

Fig 11: A detail from the 1:5 million shaded-relief map of Elysium (MC-15) by Jay Inge (1978).

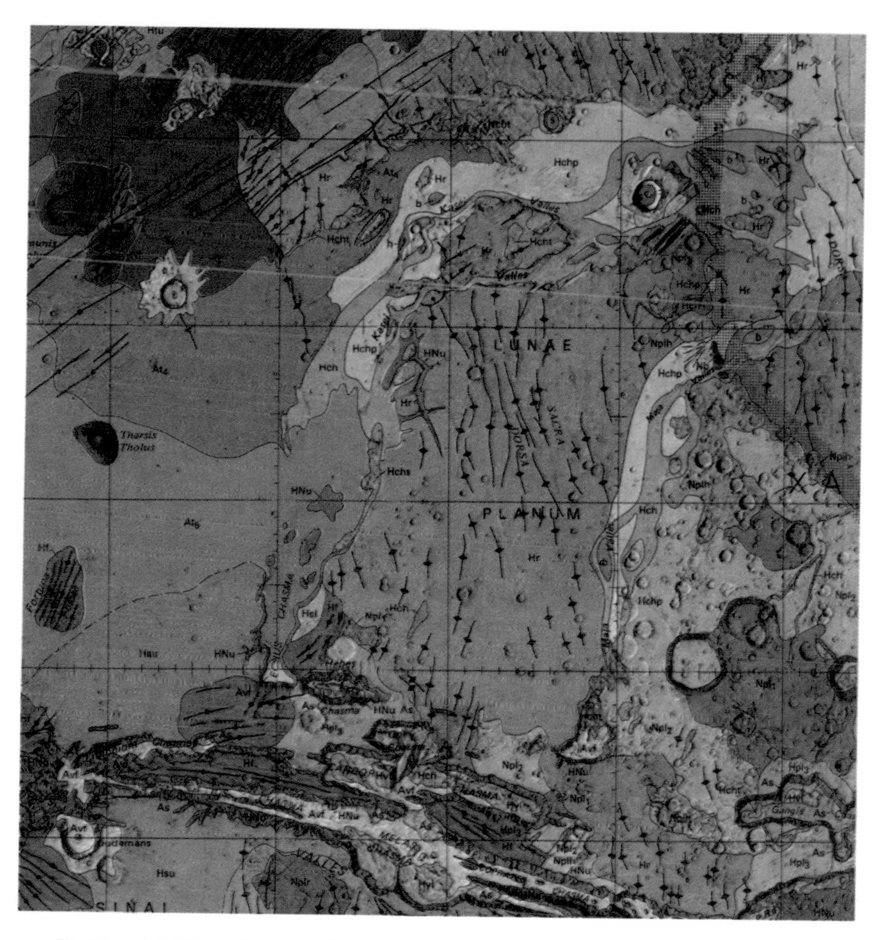

Fig 12: A detail from the geological map of the western hemisphere by David Scott and Kenneth Tanaka (1986); this shows the area north of Valles Marineris, with Lunae Planum flanked by the outflow channels of Maja Vallis (east) and Kasei Vallis (northwest).

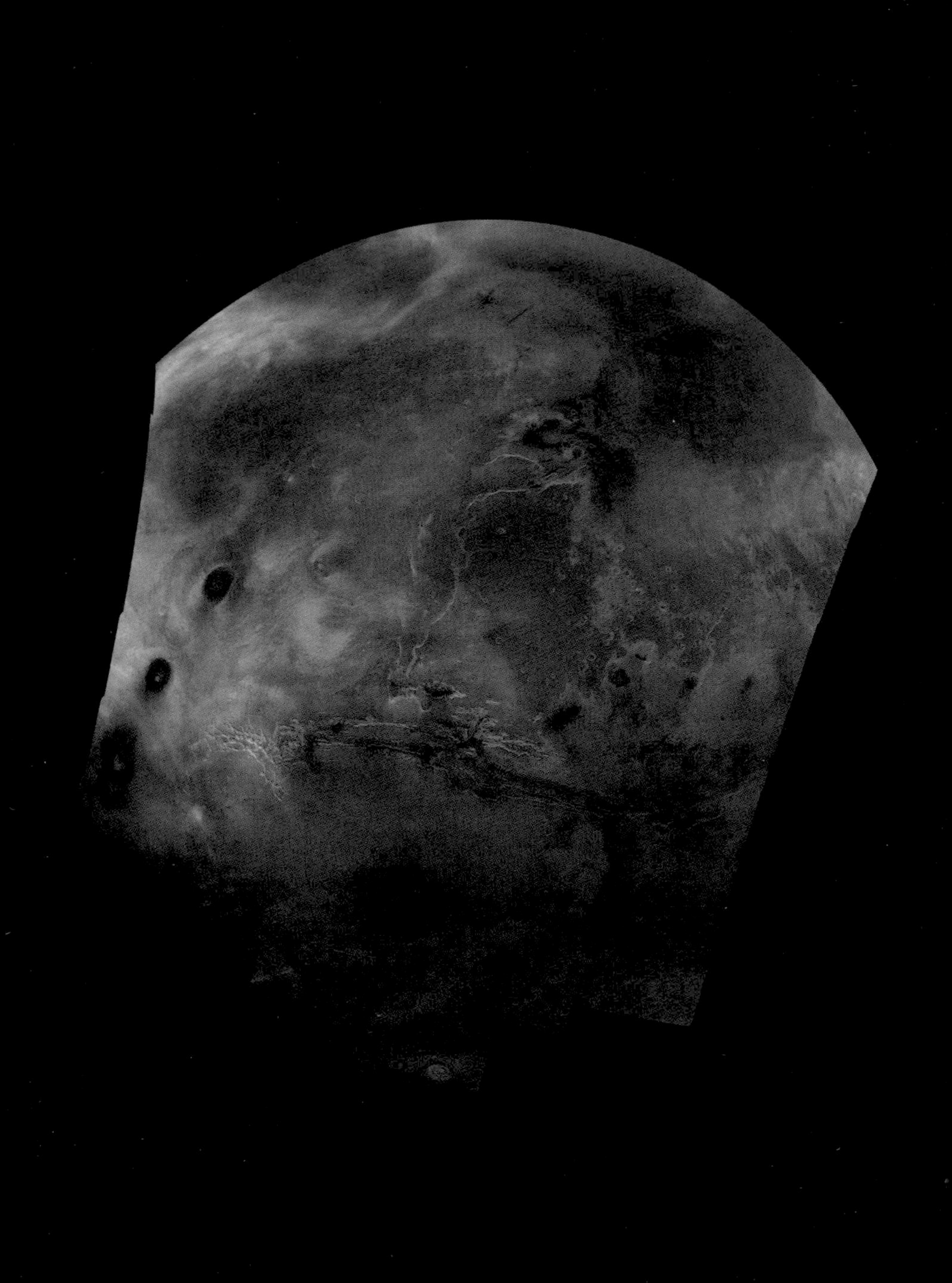

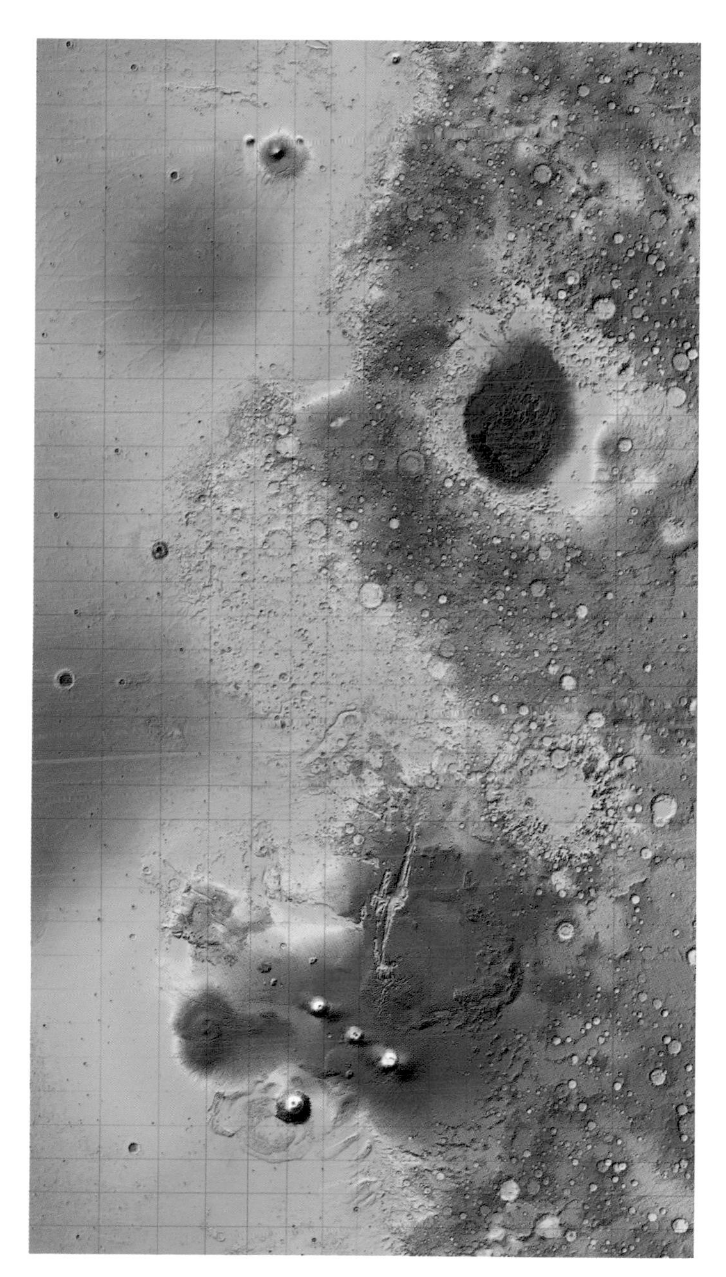

Fig 14 (above): David Smith's pride and joy: a map produced from the MOLA database (2001).

Fig 13 (opposite): A USGS *Viking*-imagery photomosaic by Alfred McEwen (late 1980s). Features from figure 12 should be recognizable north of Valles Marineris. The dark tongue in the top right is Acidalia Planitia; the Tharsis volcanoes are surrounded by cloud.

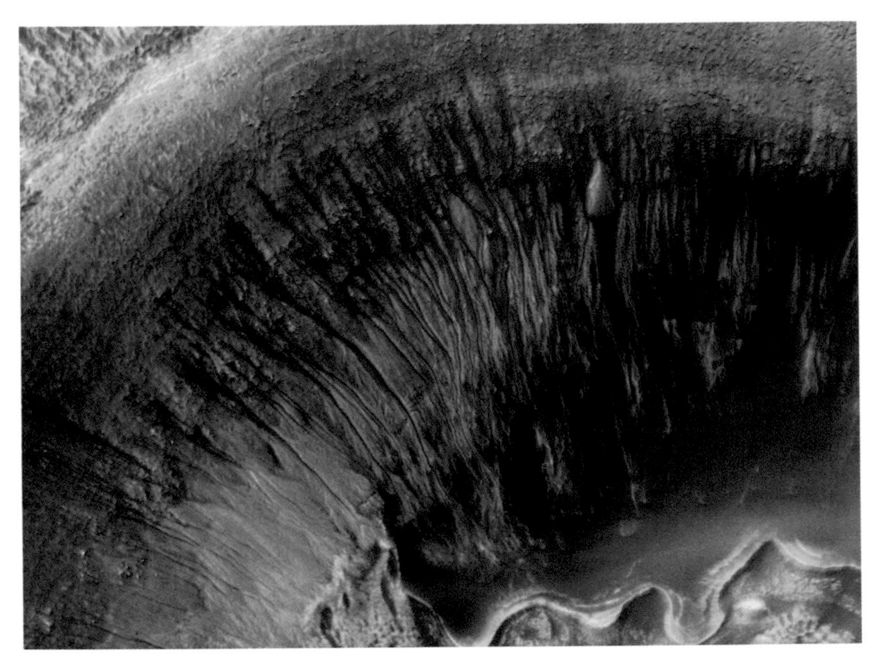

Fig 15: MOC image of gullies in a small crater within the crater Newton (2000).

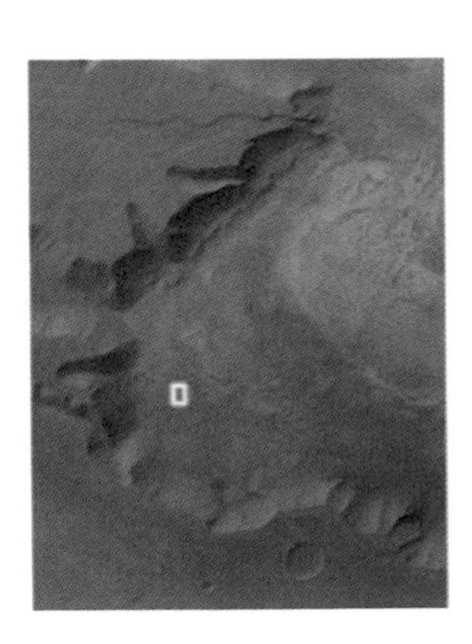

Fig 16: MOC image of western Candor Chasma (1999); the rectangle marks the site of figure 17.

Fig 17: High-resolution MOC image of stacked layers, probably sedimentary, on the floor of Candor Chasma (1999).

Fig 18: MOC image of a layered "promontory" in the crater Terby, just north of Hellas Planitia (1999).

Fig 19: A painting by Chesley Bonestell that attempts to show the weakness of Martian gravity; the temple of Zeus at Olympus is reimagined on Mars, its pillars thinned because the weight they need to carry is reduced (1956).

Fig 20: A Bonestell of the earthly temple of Zeus, for comparison.

Fig 21: William Hartmann's contemporary Martian volcanic eruption, painted in homage to Church (2000).

Fig 22: Cotopaxi, by Frederic Edwin Church (1862).

Fig 23: “First Light” by Pat Rawlings (1989).

Fig 24: Engraving of the Toroweap by William Holmes, from the *Atlas of the Tertiary History of the Grand Cañon* (1882).

Fig 25: Detail from a panorama of the Mars *Pathfinder* landing site showing the *Sojourner* rover (1997).

Fig 26: "Sand Dunes, Carson Desert, Nevada" by Timothy O'Sullivan (1877).

Fig 27: Lunar rover tracks in Palus Putredinis on the moon, with Mount Hadley in the background; photograph by James Irwin (1971), remastered by Michael Light.

Fig 28 (following page): Sequence of images showing the return of a northern ocean to a terraformed Mars. By Michael Carroll (1989).

faulting, and radial grooves suggested to the geological mappers that there might be more. For some Martian geologists the discovery of basins became a sort of cottage industry.

Grandest of all the possible new basins was Borealis, suggested in 1984 by Don Wilhelms, who had worked on *Mariner 9*, and Steve Squyres, then an up-and-coming planetary scientist at Cornell. Wilhelms has devoted relatively little of his time to Mars; he has become a moon man through and through, playing the same coordinating rule for lunar geological mapping that Dave Scott played for Mars. One of his pet lunar ideas was that a vast superbasin dominated its eastern hemisphere, a "Procellarian" basin two thousand miles across on which the familiar Imbrium basin, among others, was a later imposition. From that idea it was an easy step to the idea that Mars's northern lowlands might represent an even bigger basin. Wilhelms and Squyres suggested that a range of features in the northern hemisphere represented remnants of a basin 4,800 miles across, centered at 50° N, 170° E; it would cover the north pole itself and much of the surrounding Vastitas Borealis, as well as the provinces of Utopia, Elysium, Arcadia, and Amazonis, extending over almost 30 percent of the planet's surface. The lowlands, said Wilhelms and Squyres, had for the most part been formed in a single great blow.

Wilhelms and Squyres pointed out that, given what was generally believed about the early solar system, such a vast basin was quite plausible. Statistics suggested that if Mars had undergone a barrage of planetesimals similar to that which was recorded on the moon, it would have stood about a one in three chance of being hit by something big enough to form their Borealis. The problem with their idea was not that such an impact could not have happened. It was that the evidence for it was equivocal. The features that Wilhelms and Squyres saw as possible ramparts of Borealis were seen by other mappers as the edges of smaller (though still vast) heavily eroded basins. There were arguments for such basins in Utopia, Elysium, and north Tharsis. What was more, the dichotomy between the northern plains and the southern highlands isn't terribly circular. So however big it might have been, a single Borealis impact simply couldn't explain the northern lowlands all on its own: Extra impacts would be required around the edges for fine-tuning. Even then there would be some bits of the dichotomy unaccounted for.

Doing without Borealis, though, didn't mean abandoning an impact-based explanation. Between them, after all, proposed impact basins in Chryse, Acidalia, Utopia, and Elysium account for a lot of the lowlands. If there were a few more basins whose traces had not yet been seen, the lowlands might be a patchwork of basins Hellas-sized or a bit bigger, some already mapped, some now too far gone to be mappable. But this idea, too, has a variety of problems. For a start, not everyone believes in all of the basins invoked. The idea that Chryse is a basin is broadly accepted; MOLA measurements have shown that the plains of Utopia, recognized as a basin through delicate geological mapping, do indeed sit in a large, round depression. Other northern basins, though, are less widely accepted. Then there was the problem that the northern plains are uniformly low, while a surface that is peppered with impact basins will still have raised areas between them. Each new basin would throw out ejecta that filled in the old ones. And the idea that all the biggest impacts in Martian history would be next to each other in the north, rather than spread all over the planet, was dubious.

Given the problems with impact scenarios, what are the alternatives? Some stretch back to Gilbert himself: Perhaps the asymmetry had been there since the planet's original accretion out of objects of different densities. Another idea was that Mars's metal core might not have separated itself from the stony mantle until comparatively late in the Noachian, and that this stygian cataclysm would have led to a plume of material rising through the mantle and removing the lower part of the northern lithosphere—the hard shell that comprises both the crust and the upper, cooler parts of the mantle—thus pulling the surface down.

There is now evidence that the core in fact formed right at the beginning of the planet's history, but the idea that movement in the mantle might have simply dropped a section of the crust down a few miles retains some attractions. Among other things, it explains the conclusion from detailed mapping that some features at the boundary—mostly faults—date from the early Hesperian, while some of the features in the plains, such as the remnants of the Utopia basin's rings, date from the Noachian. During the Noachian period, proponents of the sinking plains suggest, Mars was essentially undivided, with basins and craters everywhere; then at the beginning of the Hesperian, the northern parts sank a few miles en masse. Subsequent volcanism and sedimentation would have covered all but the most prominent features in the newly low lands, but would have left faint ghosts of the past topography, like those picked up in the mapping of Utopia.

This approach, too, has its problems. The main one is that no one really knows how you would go about lowering a third of the surface of a planet like the elevator on the deck of an aircraft carrier. And if you do find a way to get some mechanism in the mantle to do this, you have to explain why that mechanism doesn't move things around from side to side as well. Or you have to accept that it could move things from side to side—in which case you have more or less worked your way around to a theory of plate tectonics, since plate tectonics deals with the way chunks of the lithosphere (plates) move around over a circulating mantle.

The fundamental insight of the 1960s' plate tectonics revolution was that convection currents deep in the mantle, a result of the transfer of heat from the planet's hot interior toward its cooler surface, are linked to processes near the Earth's surface that create new portions of the crust and destroy older ones. The creation takes place along the ridges that run along the floors of oceans. The destruction takes place when a piece of oceanic crust is overridden by another plate and sinks into the mantle. The ridge between Europe and North America churns out new oceanic crust to the east and west as the two continents move apart; subduction zones circle the Pacific, the ring of fire where oceanic crust is destroyed.

When the plate tectonic revolution was at its height, Carl Sagan, never afraid to speculate and always alert to trends, suggested in a paper with his student James Pollack that the linear features Lowell had seen as canals might in fact be the Martian equivalent of mid-ocean ridges. Once *Mariner 9* had revealed the planet's surface in its entirety, though, there seemed little room for Martian plate tectonics. Admittedly, Valles Marineris looked like a rift valley; but plate tectonics pushes continents together as often as it pulls them apart, creating compressional structures (the Alps, the Himalayas) as well as extensional ones (such as the East African rift, or the midocean ridges). On Mars, most faulting seemed to be extensional. And moving a set of rigid plates across a spherical surface necessarily cracks them so as to produce what geologists call "strike-slip" faults: Faults where one side has moved forward along the fault with respect to the other. On a pockmarked surface like that of Mars, strike-slip faulting would show up nicely, displacing one part of a crater with respect to the other. But almost no such faulting was seen: Faults on Mars all seemed to be the movement of one bit of crust up or down with respect to the other, rather than movements from side to side.

In this, as in other things, Mars looked like the moon. And that made sense, since both are a lot smaller than the Earth. Plate tectonics depend on the Earth's internal heat, mostly generated by the decay of radioactive elements in the mantle. Bigger planets have bigger mantles and thus more internal heat. It was quite reasonable to assume that the Martian mantle, seven times less massive than the Earth's and wrapped around a smaller, colder core, lacked the oomph required to push the lithosphere around.

So Martian geologists never really thought much about plate tectonics. In the late 1980s, though, someone outside the field started to get interested. Norm Sleep, a geophysicist at Stanford University, is widely respected as an expert on the way in which the Earth's mantle drives plate tectonics. When his thoughts—which range wide—turned to Mars, the northern lowlands seemed to cry out for an explanation in terms of his particular expertise. In 1994 he provided that explanation. In early Martian history, something like one of the Earth's midocean ridges had formed along the long arc of the highland-lowland boundary that runs between Isidis and Amazonis. This rift started to create new dense basaltic crust. Once new crust is being produced, old crust has to start being pushed back into the mantle at subduction zones: Sleep suggested that a subduction zone ran along the north of Tharsis and Arabia. Old cratered crust that had previously covered the high northern latitudes was pushed down into the mantle along this new subduction zone, its place taken by the new, thinner, denser crust produced at the ridge. As time went on the ridge moved north, away from the dichotomy, leaving smooth new crust behind it. (Since ridges produce crust on both their sides, they always push themselves away from the places where they begin; that's why the mid-Atlantic ridge is in the middle of the ocean.) This tectonic process only stopped when part of the ridge ran into the subduction zone along the north of Arabia; the other section of ridge ended up stuck to the western and northern flanks of Tharsis, where Olympus Mons and Alba Patera are today. Its fires, Sleep suggested, could have fueled the volcanoes' prodigious growth.

Norm Sleep's proposal did offer an explanation for the flatness of the plains, as well as their low level—on the Earth, basaltic crust produced at fast-moving ridges tends to be very flat indeed. But the flatness of the plains does not necessarily have to be explained that way; it could come from flat sediments or lavas covering an originally knobbly surface. Sleep's story was ingenious in the way it tried to account for some of the oddities of the northern plain: The stalled ridge would account for some strange outcrops north of Arabia; and a strike-slip fault in the ridge would explain an oddly straight set of hills—Phlegra Montes—northeast of Elysium. But these were not compelling readings and there were alternative explanations. Phlegra Montes could well form part of one of the outer rings of the Utopia basin (the fact that Sleep didn't spell Phlegra correctly may not have helped his case with the Mars experts). In general, the evidence for plate tectonic features in the plains was seen as slight. And again there was a real problem with the ages. The Noachian structures in the northern plains predated the faulting at the edge of the highlands that Sleep saw as evidence for the onset of rifting, and clearly the plains could not predate the creation of the rift that had made them.

In 1999, though, when *Mars Global Surveyor* began its mapping mission, plate tectonics of some sort made something of a comeback, thanks to a wealth of new geophysical data. First MOLA showed that the plains, always known to be low and relatively flat, were in fact remarkably flat; by some measures, the flattest surface in the solar system. The only comparably flat places on Earth were the ocean floors, where flattening sediments sit on already flat crust. And then the *MGS* magnetometer produced one of the mission's most profound surprises: broad stripes of magnetism across a large swathe of the southern hemisphere south and west of Tharsis.

Earlier Mars missions had detected barely a trace of a magnetic field and this was not a great surprise. The Earth's magnetic field is generated by currents in the liquid outer layer of its core. Mars is not thought to have a hot liquid layer in its core. Again, the problem is a lack of heat. Mars, like the other rocky planets, would have been born hot—a fearful amount of energy is released as the constituent planetesimals smash into each other—but, being small, it would have lost that heat quite quickly. Its iron core would quickly have frozen solid, at which point any currents generating magnetic fields would have ceased. The magnetic field detected by *MGS*, though, was not coming from the core: It was coming from the rocks in the crust. Igneous rocks contain minerals that can act as tiny magnets and, when they cool below a certain point, these tiny magnets align themselves with the local magnetic field. From then on that's the orientation they are stuck with. Part of the plate tectonic revolution was an understanding of what these magnetic fields frozen into rocks could reveal about the past. The fields detected on Mars seem to be similar remnants of a bygone age: Sometime in the early Noachian, when the crust of the southern highlands cooled down enough to have a magnetic field imprinted on it, there was a magnetic field there to do the imprinting.

It was not just the fact of the magnetism that came as a surprise. Its strength and pattern were also remarkable. The remnant magnetism was much stronger than was ever seen on Earth. And the magnetic field was arranged in long parallel stripes—which made Norm Sleep very excited. Vast magnetic stripes along the floor of the Atlantic had been one of the keys to plate tectonics. The new crust produced at ocean ridges preserves the magnetic field at the time that it cooled down, so since the direction of the Earth's magnetic field flops over on a fairly regular basis the magnetic polarity of the rocks on the ocean floor changes regularly: Looked at in terms of magnetic field, the floor of the Atlantic is covered in stripes of opposite polarity arranged symmetrically around the central ridge.

Stripes in the Martian crust suggested it might have been made in a similar way.

As always, there are other explanations. If you take a bar magnet with a north pole and a south pole and break it in two, you get two magnets each with a north pole and a south pole; the more you break them, the more poles you get. Similarly, if you take a magnetized lump of crust and fracture it with faults, you'll break up its uniform field and make it stripy. But even if these stripes are not produced in the same way as the stripes along earthly ocean floors, they still speak to a very powerful primary magnetic field, and thus an active interior that makes plate tectonics more plausible than they would otherwise be.

Another set of data from *MGS* has also been taken as evidence that internal, rather than external, effects created the northern plains. A planet's gravitational field is not uniform: If you stand close to a big mountain with a plumb line in your hand, the bob will be attracted toward the mountain and the line will not point quite toward the center of the Earth. In the nineteenth century, though, surveyors in India noticed that their bobs were not as attracted toward the Himalayas as they would have expected. An explanation for this effect was offered by the Astronomer Royal, George Airy—a nonastronomical insight that was his greatest contribution to science. Airy's suggestion was that the crust of the Earth was thicker under the mountains. Assuming that the crust floats on the mantle beneath it because it is lighter, thicker crust will both sink farther into the mantle and stick up farther above it (think of a thick log floating next to a thin plank). Thus mountains must have deep roots that are less dense than the surrounding mantle that they float in—and this relative lack of density down below accounts for the smaller than expected effect of mountain ranges on the gravitational field at the surface. (The same principle applies to the ocean floors: The crust created at ocean ridges is thin and dense, and so it floats lower in the mantle than the thicker, less dense continents.) This floating is known as isostatic equilibrium. Among Gilbert's great contributions to geology were his measurements of the way the crust in the Utah desert had risen as it sought to reestablish isostatic equilibrium when relieved of the burden of a large lake.

From the beginning of the space age it had been clear that accurately tracked spacecraft provide a very good way of measuring gravitational fields. In the early 1960s America's Transit navigation satellites allowed an unprecedentedly accurate map of the oddities in the Earth's field to be made—information vital if America was to get its ballistic missiles to fall in more or less the places it was aiming at. Then JPL's Lunar Orbiter spacecraft found that some lunar maria were more attractive than they should be; they represented concentrations of mass, "mascons" in the jargon, powerful enough to make quite a change to a spacecraft's orbit. The explanation for this was that the moon, being small and thus cold, has a very thick lithosphere that can support great weights. When thick basaltic lavas filled up large impact basins, the floors of the basins did not sink to isostatic equilibrium: The thick lunar lithosphere simply supported the lavas where they were, producing the unusual mascons.

MGS has provided the best data yet seen on the Martian gravitational field. Much of the planet seems to be in something like equilibrium. The Tharsis bulge stands out as a positive anomaly, suggesting that its height is supported by a very thick, strong lithosphere, rather than by the buoyancy that would be provided by peculiar deep crustal roots. Two of the obvious basins—Argyre and Isidis—show mascons like those seen on the moon. So does the Utopia basin in the northern lowlands, suggesting that it has been filled in by lavas or sediments many miles deep. But though Utopia looks like a basin in this way, the northern plains as a whole do not. The gravity data offer no obvious support for either the one-big-impact idea or the clustered impact idea, indeed, it seems to argue against them.

Combining these gravity measurements with the detailed topographic data provided by MOLA has allowed *MGS* scientists to make a rough estimate of how the Martian crust differs in thickness around the planet. Differences in crustal thickness suggest that the flow of heat through the southern highlands in the earliest days of the crust was lower than one would expect, given what is thought to be known about the total supply of internal heat available. Less heat in the south suggests more heat in the north, pumped from the deeps by some sort of convection in the mantle. Mantle convection like that which, on Earth, drives plate tectonics.

So, after thirty years of mapping and study, the most salient geological feature on the face of the planet lacks an agreed explanation. The multiple working hypotheses all have their problems. More tests are needed, which means more data. Most of all, especially from the geologists' point of view, dates are needed. Solid, reliable dates by which their threefold history of Mars could be tied down, which could say for sure when the dichotomy formed, and which could help answer the other great question raised by the idea that movements in the mantle have shaped Martian history. Have those movements stopped completely? Or does Mars still have enough internal heat to drive a little activity? Are its volcanoes long dead, or merely dormant?

The Artist's Eye

> One of the foremost tasks of art has always been the creation of a demand which could be fully satisfied only later.
>
> —Walter Benjamin, "The Work of Art in the Age of Mechanical Reproduction"

At the end of 2000, Bill Hartmann's answer to the question of Mars's inner activity sits on an easel in the middle of his studio, a few blocks away from the University of Arizona, Tucson. A dark volcano climbs gently from a rusty plain. Behind it, the sky changes color. Near the horizon it is dusty red; toward the zenith it is a dark, bewitching blue, almost the blue of an Arizona evening. Here, though, the blue is a passing shade of twilight—by day the sky would be washed yellow by fine dust from the plain. Now, darkening, it is split by clouds of smoke from the mountain's peak. The sun catches the clouds from beyond the horizon; an angrier light, hot and red, lends fire to them from below.

Hartmann has been studying Mars since *Mariner 9* and he is one of the world's undoubted experts on the age of the Martian surface. He's also an experienced landscape photographer and an artist of talent and some modest fame (not to mention a writer and novelist as well). The common theme in his achievements is a love of landscape—earthly and unearthly. As a child he sketched and painted the world around him; in the evenings he looked through his telescope at other worlds. As a teenager in the 1950s he used his telescope to study and map the moon. Then he built plaster of Paris models of his favorite crater, Walter, to try to appreciate what it might be like from ground level, to turn a planetary object into a place. He wanted to answer the question that has driven him ever since: To see what it would be like to be there. The erupting volcano in his latest painting is an answer to that question; an answer based on his ever-strengthening belief that Mars is still an active planet.

It was this desire to get a better point of view that led to Hartmann's first major contribution to science. After majoring in physics at Penn State, he moved to Tucson to take a masters in geology and a doctorate in astronomy as part of the new planetary science program that Gerard Kuiper was starting at the University of Arizona. Kuiper had struck upon a new way of seeing his moon and set Hartmann to work on it. The new technique was to take a high-resolution transparency of the full moon and project it, not onto a flat screen, but onto a white sphere. Features normally foreshortened spread themselves out over the surface of the sphere as it sloped away, and by wandering from side to side it was possible to look at the familiar face of the moon from new angles. Using this new unearthly perspective, Hartmann saw aspects of the surface never before appreciated. Most strikingly, Mare Orientale, an obscure dark patch forever on the limb of the moon as seen from Earth, was revealed as the center of a startling multi-ringed bull's-eye of an impact basin, perhaps the most dramatic single feature the moon's face has to offer.

Impacts went on to become a major theme in Hartmann's work. In the 1980s he was one of the researchers responsible for the now widely accepted theory that the moon was actually formed in an impact—one to dwarf even the impact invoked to explain the Borealis basin. The idea is that as planetesimals accumulated to form planets in the very earliest days of the solar system, they produced not four inner planets, but five. The fifth planet was about the size of Mars and in an orbit that crossed the Earth's. It and the Earth eventually collided. The planets' cores merged; their mantles and crusts melted and mixed; and some of the ejecta was thrown out into space. Orbiting around the molten Earth this ejecta coalesced to form the moon. This theory has some nagging inconsistencies, but in general it explains well the ways in which moon rocks differ from rocks on Earth and the ways in which they are the same.

Bill Hartmann is not the first artist to include the face of the moon when he paints or photographs landscapes. Jan van Eyck's detailed representations of the moon in a pair of sixteenth-century altarpieces mark a watershed in the realistic portrayal of the natural world. The American West, where the moon seems so oddly present, has seen much use of the same idea: Think of Ansel Adams's *Moon over Half Dome,* or indeed of *Moonrise, Hernandez, New Mexico*, perhaps the best-known photographic landscape of the twentieth century. There, as in the van Eyck crucifixion, the moon offers a lifeless counterpoint to the crosses in the foreground. Hartmann is not an artist in that league, but when he includes the moon in one of his earthly landscapes he has the joy of knowing where it comes from, a subject he is happy to talk about with prospective purchasers. "It seems to me that people might like a landscape painting with a daylight moon by 'the person who helped figure out the origin of the moon,'" he told me with a laugh when I visited his studio. "I figure this effect will kick in as soon as I die and it's too late to do me any good."

While taking his degrees, Hartmann spent little time on art. In the mid-1960s, though, when he was commissioned to write a textbook, he saw that including line drawings would make it much more interesting—textbooks were dry in those days and photo reproduction an expense academic publishers found easy to forgo. Still, he didn't go out of his way to make his colleagues in the sciences aware that he painted planets as well as studied them, for fear they would look down their noses at him. "It was like being in the closet," he says. In the 1970s he put together a popular book about the discoveries made through *Mariner 9,* called *The New Mars.* It includes a few paintings by others—the Czech artist and author Ludwig Pesek, Don Davis, who worked with the Hayden planetarium, and the great Chesley Bonestell, whose *Life* magazine pictures of astronomical scenes had adorned Hartmann's walls when he was a boy—but none of his own.

Slowly, though, he started to out himself to his planetary-science colleagues. "I'd listen to a talk and do a painting and send a slide to the guy who'd given the talk." After a while he began to see the slides turning up in further talks; not long after scientists started to suggest new topics for paintings. In 1981 he and Ron Miller, another author and artist, produced a book called *The Traveler's Guide to the Solar System* featuring paintings of all the major planets and many of their moons, not to mention assorted asteroids and a comet or two. There are images produced by spacecraft in the book, too, but the paintings take pride of place. The pictures are scrupulously honest, trying to imagine nothing that cannot be backed up by appeal to the facts. At the same time they are gloriously dramatic. Jupiter and Saturn hang above the surfaces of their moons in implausible splendor, echoing but far surpassing the moon in the Earth's sky. The great wall of Mare Orientale glows in glorious blue earthlight.

The Traveler's Guide, revised and expanded in the 1990s as *The Grand Tour,* is probably the best-known showcase of realistic extraterrestrial landscape art around. The genre is a small one; there are only about a dozen artists in it that any but the most die-hard aficionado could name. Kim Poor's Novagraphics gallery in Tucson is the only commercial outlet devoted to the stuff (and even there, most of the money comes from signed Apollo memorabilia). It may seem strange that this profession should persist at all in an age when the planets have been visited and photographed in close-up. But it is precisely because of planetary exploration that the paintings are both possible and popular. The artists, while working in a strictly realist idiom, can still create landscape in a way that the controllers of a spacecraft camera cannot. As Kim Poor says, "Artists go where engineers fear to tread." Reproduced at planetariums as posters and in books and magazines, the aesthetic these artists are committed to still has a powerful influence on the way that the solar system is seen.

In some ways the art has an oddly old-fashioned feel. It is art that prizes representation very highly, which minimizes the role of brushwork and gesture. It has a close relationship to technical illustration and prizes the mimetic over the expressive. It is not that the artists are blind to other approaches. In Hartmann's studio there's a striking work in progress that he began at Artist's Point, the place where Thomas Moran painted *The Grand Canyon of the Yellowstone* in the 1890s. In tribute to Moran—and Moran's inspiration, J. M. W. Turner—the picture has a strong sense of movement and expression, of light as something solid. But such effects are largely missing from Hartmann's astronomical paintings, as from those of his peers.

Richard Poss, who teaches at the University of Arizona, argues that today's astronomical art is usefully understood in the context of the school of American landscape art that Moran was extending and moving away from, the Hudson River School of Thomas Cole and Frederic Church. The Hudson River School, as the critic Robert Hughes has written, played a key role in turning America from territory to landscape, from a thing of exploitation to a thing of contemplation. Their landscapes celebrated the expansiveness of the American scene with a clarity of vision that offered detail and drama. The specific topography was not necessarily important: The thing that mattered was communicating the possibility of a new world and bringing its brightness directly to the eye, which is just what Hartmann and his peers try to do.

Unlike Turner and the landscape artists who followed him, for whom expressing the experience of the landscape became so important, the astronomical artists choose largely to limit themselves to what landscapes should look like, rather than how they might feel. A feeling would be speculative; the look can be ascertained with some degree of accuracy. And just as the artist is to some extent excluded from the work, so, often, is the human form. Perhaps the most influential astronomical paintings of all time were Chesley Bonestell's first portfolio of Saturn as seen from its various moons, published in *Life* magazine in the 1940s. In these amazing pictures there is simply no place for people. In a way this comes as no surprise. Bonestell had worked as an architect (playing a role in the design of both the Chrysler Building and the Golden Gate Bridge), and as a matte artist in Hollywood, painting backgrounds that would trick the eye into accepting as real places the camera could not actually capture, such as the palatial Xanadu he created for Orson Welles's *Citizen Kane.* Like an architect, the matte painter creates a world for others to inhabit.

Bonestell soon found himself putting people in the pictures, though, as when he collaborated with Wernher von Braun to illustrate the human exploration of Mars for the readers of *Collier's* magazine. As a commercial project, planetary art with people in it has a much larger market than pure landscape. The subject is changed from observation to travel and the painting becomes a piece of science fiction, rather than astronomy. While Hartmann often does without people completely (there are hardly any people in *The Grand Tour,* even in the pictures of Earth), he sometimes adds them to an otherwise finished painting. Sometimes these people—whom Hartmann normally paints in blue space suits, in part because "our eyes on Mars will be starved of blue"—help the composition. Sometimes, I suspect, they are there for the simple reason that Hartmann wishes he could be. And sometimes they are the result of a commercial calculation. There are private fans and corporate sponsors—big aerospace companies and NASA, for the most part—who make such science fiction worth painting.

Some sensibilities see the frequently commercial transition to science fiction as a diminishment; others see it as a virtue. It certainly adds to the art's potential, allowing it to express a wider range of ideas. Poss sees the way today's astronomical artists use people as an intriguing ideological break with the legacy of the Hudson River School. Thomas Cole and his cohorts were engaged in a reactionary project. They recorded the Hudson and the wildernesses to the west not as they were at the time, but as they had been before the beginnings of industrialization—the process that provided their patrons' wealth. While man appears in their wilderness images, he is either a lone icon—Daniel Boone by the stream—or a shadowy axe man, waiting to fell the forest. In today's art of planetary landscapes, the presence of people is, by contrast, generally positive. They are studying the landscape with awe and respect; they are building a future. Such paintings combine a romantic respect for nature with a technophilic optimism. They are allegories of progress that embody a rare and pleasing belief that the good can come to, rather than just from, the beautiful.

Mars lends itself to such treatment particularly well. In large part this reflects the fact that Mars is a potential site of manned exploration; people on Mars have a far more immediate iconic significance than people on obscure bits of ice at the far end of the solar system. Another factor may be that the focus on the sky that is often the hallmark of purely astronomical art—such as the Bonestells of Saturn from its moons—doesn't really work on Mars, with its hazy atmosphere and frankly disappointing moons. Many more artists have painted Mars from its moons than have painted the moons from Mars, because from the moons the view is quite superb: Mars broad and gaudy in the vacuum-clear sky.*

The typical Mars image is of spacesuited explorers—normally a team of them—in a somewhat Arizona-like landscape, often with dust clouds in the distance, like the storm clouds of the Hudson Valley. The explorers are in some situation of discovery; they are finding a geological oddity, or a fossil, or a spacecraft from a bygone era, such as ours. Normally it is something meaningful, occasionally it is something allegorical. In *The Key,* by Pat Rawlings, a globe of the Earth lies nestled in the Martian sand. There is always a sense of temporal tension, of the land and its secrets being old and the explorers new. In *The Key* the continents on the globe are arranged as they were 250 million years ago, and yet this ancient artifact is covered by just a little windblown dust.

Mars, these images say, is old and changeless, but made new again by those who see it for the first time. Perhaps the best-known painting of such a discovery is *First Light*, also by Pat Rawlings, in which the object of discovery is Mars itself. An astronaut is silhouetted at the edge of a canyon, setting up an anemometer, while a companion lowers himself down the cliff face; behind them the sun is rising and the mist in the void below them captures the shadows of the canyon's far wall. The cliff in the foreground and the sky far from the sun are dark as night, the mists glow silver, the sunrise golden. If there's an antecedent to this image it's not among the precise renditions of Cole and Church. It is Caspar David Friedrich's romantic masterpiece *Wanderer Above the Sea of Mist.* But whereas in Friedrich's painting the backlit figure looking out over the bright

*There is one very striking piece of art that takes one of the moons as its subject, but it is a photograph, not a painting. Ed Strickland is a Mars enthusiast who, but for graduate work burn-out, would have become some sort of planetary scientist, and who now works as a computer consultant in Texas. Starting with an image of Phobos taken by the *MGS* MOC, he screened out the sunlit part of the moon and enhanced the brightness and contrast on what was left. The ghostly result is the face of Phobos lit by Mars-shine. Quite why this is so impressive I am not sure: If you saw it without knowing what it was you'd just think it was a rather poor picture. Perhaps it is the fact that this is the only picture I have ever seen in which Mars is actually doing something—taking an active role in the universe. Perhaps it is just the fact that it was a neat thing to think of doing.

mountain landscape is a central intrusion, blocking our view as he contemplates nature, in Rawlings's piece the two figures are smaller and off-center. They are within the landscape, not above it. Only the standing figure's anemometer mast breaks the horizon, its horizontal vanes offering a faint evocation of some alpine Calgary. Rather than simply looking at the sublime, they are studying it, measuring it, lowering themselves into it. Friedrich's rambler has arrived; Rawlings's astronauts have a world still to discover.

One purpose of such figures in landscapes is to offer a sense of scale. On Mars, though, this rarely works. The most spectacular attributes of Mars—the vast canyons, basins, and mountains—cannot be captured because they are simply too big. The canyon lands of *First Light* look very modest by Martian standards, probably a minor part of the Noctis Labyrinthus system; the detail that can be seen on the far side shows that they have nothing like the scale of Valles Marineris, though many who look at the picture doubtless think that is what it represents. As for the mountains . . . "Everyone asks, 'Did you do a painting of Olympus Mons?'" says Hartmann with a laugh. "Well, you can't. If you back off far enough to see it you're too far away. It's the size of Missouri. It's like a picture of England as seen from France. You can't paint the Rocky Mountains in one painting." The artist's dilemma in trying to capture planetary phenomena on a personal scale is brilliantly captured in "The Difficulties Involved in Photographing Nix Olympica," a short story by the British author Brian Aldiss. Its focus is not on the photographer himself, but on the acute mixture of alienation and agoraphobia felt by the companion he sends off into the barren landscape in a vain attempt to add some sense of scale to the all but imperceptible rise of the mountain.

Aldiss realized that the most fruitful approach to the Martian landscape would not be to try to understand it on the human scale, but rather to contrast it to the human experience. Perhaps the first truly successful use of the planet to this end was in *Watchmen*, an ambitious and accomplished graphic novel written by Alan Moore, drawn by Dave Gibbons, and published, like Aldiss's story, in the mid-1980s. One of the principal characters in *Watchmen* is Jon, a once human superhero whose vast powers over time, space, and the structure of matter have made relating to humanity hard for him;

reasonably early on in the action he removes himself from the Earth. Gibbons, looking for inspiration, came across *The Traveler's Guide* in a library and was captivated by its chapter on Mars. He loved the realistic treatment it offered of an alien, inhuman world; he was also struck by some strange synchronicities. Most extraordinary was seeing a picture of the smiley face in Galle crater; extraordinary because a smiley face (with a splash of blood across it) was a key part of the graphic novel's recurring imagery. He enthused to Moore about the possibilities these Martian landscapes offered. As a result the novel's ninth installment sees Jon and his one-time lover, Laurie, floating over the planet's best-known landmarks as they talk about the most intimate details of Laurie's past and the nuclear apocalypse threatening the Earth. Godlike Jon appreciates the vast scale and age of the landscape below them in ways that no human could—and attaches little significance to Laurie's memories or to the end of life on earth.

The failure to find any trace of life on Mars in the 1970s was as harsh a blow to science fiction as it was to science. It had almost always been the Martians, rather than their planet, on which the fiction had focused. From the mid-1970s to the mid-1980s there was remarkably little new science fiction about Mars. Inspired by *The Traveler's Guide,* Moore found a way to reclaim what had been lost by giving significance to the planet itself, rather than its inhabitants. Mars offered him a contrast to the pettiness of Earth as sharp as the divide between one panel and the next. It provided a place of timelessness to frame the sharp cuts between different events in Laurie's memory. It provided a way to talk about the absence of life as something other than death. Life, Jon tells Laurie, "is a highly overrated phenomenon. Mars gets along perfectly well without so much as a microorganism. See: There's the South Pole beneath us now. No life. No life at all, but giant steps, ninety feet high, scoured by dust and wind into a constantly changing topographical map, flowing and shifting round the pole in ripples ten thousand years wide. Tell me—would it be greatly improved by an oil pipeline?"

Moore made memorable use of Mars, and Gibbons got the opportunity to create his own renditions of the landscapes he had discovered in Hartmann's and Miller's book. But he also found himself having to try things Hartmann and Miller had wisely avoided. Moore devoted a page of the script from which Gibbons worked to building up Olympus Mons, "A sizeable mountain, very far away . . . The sizeable mountain is now quite a large mountain, still very far away . . . The mountain is now a bloody enormous mountain, and it's still a long way away . . . Olympus Mons, now completely filling the background. It is still some distance away. We are starting to understand how incredibly huge it really is." Gibbons took his best shot at turning these instructions into images for the readers, but it defeated him, as it had to. Comic books are drawn at Laurie's scale, not Jon's.

The volcano now standing on the easel in Hartmann's studio is a much more manageable thing—perhaps the size of Etna, or maybe Ararat, and lower in profile. There are no people in the picture—yet—but there's a sort of life. The planet itself is active; its surface is renewing itself. Ten years ago, with or without people in the frame, this would have been widely seen as a fiction. Mars was ancient and inactive. Today, Hartmann is convinced that it's on the other side of the divide: that the eruption counts as realistic astronomical art, because the surface of Mars is still being renewed by volcanism.

Hartmann's approach to dating the surface of Mars works by comparing it with the moon. On the moon, it's possible to translate the ages derived from crater counting directly to absolute age measurements, since the Apollo missions brought back rocks that could be dated through the analysis of radioactive decays. If Mars had been hit by objects similar to those that hit the moon, and at the same rate, then the rules used for translating lunar crater counts into absolute ages could be applied to Mars. And even if the rate at which Mars was hit was different, a similar calculation could still be made, as long as the difference in cratering rates was taken into account. Thus if Mars was hit twice as often as the moon, a Martian surface with the same crater count as a lunar surface would be half the age of that corresponding surface, because craters on Mars accumulated twice as quickly. To stir up audiences at scientific meetings Hartmann sometimes calls the ratio between the cratering rate on Mars and on the moon "the most important number in the solar system" because it is so crucial to dating the Martian surface. And it lives up to its importance by being an infuriatingly hard number to ascertain.

Mars is closer to the asteroid belt than the moon is, which means you might expect it to get hit by asteroids more often. Because it's farther from the sun, though, it's less likely to be hit by a comet, and comets cause craters too. Mars moves more slowly than the moon and slower impacts mean smaller craters. But since it is larger than the moon, its gravity will speed an impactor up more as it gets closer. Having worked his way through all these issues, in 1977 Hartmann came to the conclusion that the rate at which craters formed on Mars was about twice that at which they formed on the moon. That allowed him to put rough dates to the three Martian geological periods. The Noachian–Hesperian boundary was put at three and a half billion years ago. The Amazonian period began 1.8 billion years ago and the surface of Olympus Mons, as fresh as anything on the planet, was just two hundred million years old.

To earthly geologists this made most of Mars terribly old: In 1977 everything that happened before the beginning of the Earth's Cambrian era, about 540 million years ago—the point when animals developed shells that could be fossilized—was considered profoundly ancient. But among planetary scientists, used to a moon where most things had happened more than three billion years ago, Hartmann's chronology soon came to be known for the youthfulness it conferred on Mars. Gerhard Neukum and Donald Wise, working with different assumptions about the best way of measuring the rate of cratering on the moon, suggested that craters of a given size were actually produced on Mars at only a quarter of the rate they were produced on the moon. Given these assumptions, Martian history looked very different: In the Neukum and Wise model, the Hesperian period ended before it had even begun in the Hartmann chronology. Almost all the volcanic activity Mars had ever seen was held to have taken place in the first billion years of its existence; by the time the solar system was half its current age, all Mars's volcanoes had shut down. Mars today was cold and dead.

For most of the 1980s the two approaches were in a standoff, though Neukum did modify his estimates a little. But evidence from a completely different field was to begin tilting things toward Hartmann's views. In the early 1980s it started to become apparent that there were already some samples of Mars available for analysis on Earth. This evidence came from eight unusual meteorites, known as the SNCs. The SNC meteorites were made of basalt, which argued that they had been knocked off a body that had a chemically distinct crust and mantle (basalt is what you get from melting and distilling the rocks of a mantle). At least one of the larger asteroids—Vesta—seems to have basalt on its surface, but the SNC meteorites did not seem to come from Vesta. Nor did they come from the moon (though some basalt meteorites do). By the end of the 1970s, Mars seemed, in the careful words of Mike Carr at the USGS, the "least improbable" source. Then researchers looking at shocked glass within one of the rocks found that it held traces of an atmosphere, in the form of unreactive ("noble") gases; and the mix of noble gases was more or less identical to the mix measured by the *Viking* landers. The meteorites had come from Mars.

Despite this evidence it took some time for the SNC meteorites to be taken seriously as evidence for what Mars might be like. Part of the problem was finding a plausible means by which they could have been launched from the planet's surface. This was sorted out a few years later, when it was argued that the atmospheric shock wave following a larger meteorite impact could boost some of the impact ejecta into orbit. By the late 1980s, then, the SNC rocks were thoroughly accepted as far-flung bits of Mars. Their chemistry provided proof that Mars had differentiated into a core, mantle, and crust at the very beginning of its history, thus ruling out the "late core formation" hypothesis for the great north-south dichotomy. Radiological dating showed that they were young. The Nakhlites (named after the first stone of the class to be discovered, the Nakhla meteorite that fell in Egypt in 1911) had cooled from their molten form about 1.3 billion years ago. The Shergottites, named after the meteorite that fell near Shergotty, in India, in 1865 were about 170 million years old. So Mars was still producing basaltic lavas relatively recently.

Quite how recently Hartmann was not able to say until the late 1990s, when MOC, the high-powered camera on *MGS,* started sending back images of the planet's surface. Hartmann concentrated on looking at the new images of areas that were already known to be comparatively young. The areas he was looking at had few large craters (that's how they were known to be young); the new images showed them as being nearly devoid of small craters, too. That pulled the age limits ever more firmly toward the present. For some surfaces in Elysium and Amazonis—and high on the slopes of Olympus Mons—Hartmann came up with ages of between ten million and a hundred million years. Alfred McEwen, at the University of Arizona, looked at the pictures and found what seemed to be exquisitely fresh textures on a great flood of lava that covers southern Elysium and much of Amazonis. All sorts of fine details were visible; again, the craters suggested an age of as little as ten million years.

There are complications. In some places, it appears that a layer of rock that lay above some of these lavas has been eroded away. If that layer protected the rock below from cratering, the lava flows might be older than their fresh face suggests. But, Hartmann asks, can this argument really apply to all the freshest surfaces—which total about four million square miles. Can it apply to the utterly uncratered areas on the high flanks of Olympus? Better, he suggests, to avoid the special pleading and go with what the landscape seems to be saying: that in some places Mars, though ancient, is still capable of renewing itself.

Layers

No! There's the land. (Have you seen it?)
It's the cussedest land that I know,
From the big, dizzy mountains that screen it
To the deep, deathlike valleys below.
Some say God was tired when He made it;
Some say it's a fine land to shun;
Maybe; but there's some as would trade it
For no land on earth—and I'm one.

—Robert W. Service,
"The Spell of the Yukon"

While Hartmann used the new MOC pictures to improve crater dating, others were using the camera to look at the other great geological record of time: layering. All planets are layered, almost by definition—anything big enough for gravity to pull into a ball is probably hot enough inside to separate into layers, which is why planets have cores and mantles and crusts. Planets with surfaces prey to volcanism and erosion will be layered at a finer scale too. But these commonplaces did not stop the sheer amount and diversity of layering revealed by MOC from coming as a surprise to everyone, not least the scientist who ran it, Mike Malin.

When Malin was a graduate student of Bruce Murray's in the 1970s he became interested in two things: Mars and how to make images of it. When plans for the *Mars Observer* mission started to be formulated in the early 1980s, Malin was the man who campaigned hardest for a camera to be included on board, an idea the geophysicists, geochemists, and climatologists who were supporting the mission didn't much care for. The core of his argument lay in the design of the camera he was proposing: He strove to make it cheap enough, small enough, and meager enough in its requirements for power that refusing it room on the spacecraft would look churlish. The secret to his success was a camera with just a single line of light detectors and no shutters: As the satellite moves over Mars the line of detectors builds up images in a long thin strip. Malin went beyond designing the camera: He built it, using a MacArthur Foundation "genius grant" to set up a company dedicated to cheap, lightweight cameras for space missions. It's headquartered in an office building outside San Diego, blank-windowed and anonymous enough to be the last place where you might expect to find a doorway to another planet. Unless, that is, you are a conspiracy theorist convinced that the government is trying to put everyone off the trail of the civilization that built the "Face on Mars" by doctoring pictures to make the Face look like just another eroded mesa on the edge of Chryse. In that case, by a curious inverted logic, it's exactly the sort of off-puttingly anonymous place you would expect.

Malin's associate in running the camera and interpreting its data is Kenneth Edgett, a younger geologist whom Malin met when he was teaching at Arizona State University. Edgett is a likeable man with something of a put-upon, harassed air, which probably comes from the simultaneously exciting, taxing, and frustrating nature of his job. MOC provides stunning new pictures of Mars on a daily basis, and especially in the first years of its operation Malin and Edgett barely had time to catalogue them, let alone to stop and think about them seriously. While other researchers champed at the bit for access to the MOC images ("validated" MOC data are released in chunks after a six-month delay, an arrangement that dates back to the original planning for *Mars Observer*) and resented the privileged access of the MOC insiders, Edgett would complain that it took all his time just to control the fire hose; drinking from it was an impossibility, especially since interpreting the images wasn't easy. Mike Carr, who had run the cameras on the *Viking* orbiters and who, like Hartmann, was part of Malin's team on MOC, frequently found the pictures bafflingly hard to interpret even when he was able to compare them directly with *Viking* pictures of the same locations. The new level of detail—MOC could pick out individual boulders—just made it all look completely different. Features that had seemed well defined before suddenly weren't; features unhinted at in the *Viking* images appeared out of the blue.

Even before everyone's eyes got attuned to the new way of seeing things, though, the first images from MOC showed that layering was going to be a major part of the Martian story. Everywhere they looked, Malin and Edgett saw layers. In crater rims, in valley walls, in mesas, on flat surfaces scoured by wind: Anywhere where the bedrock was visible there was layering. And these weren't just layers and layers of identical lavas; there were differences between the layers. To Edgett's and Malin's excitement it began to look increasingly likely that some layers—those in the smooth terrain of Terra Meridiani, for example—were undoubtedly sedimentary, though whether laid down in water or by the wind was not immediately clear. Either way, there were stories there. There was a history of change, of rocks laid down and eroded away, unlike anything that had been seen on Mars before.

If Hartmann's use of MOC for crater dating was extending Martian history right up until the present day, the images of layering were pushing its beginnings ever further back into the past. In fact, *MGS* was making the earliest years of Martian history look increasingly interesting, what with an unexpectedly strong magnetic field and unknown levels of inner turmoil. Just as MOLA was showing that there were the ghosts of older craters under the craters previously deemed to be the oldest, so MOC was showing that there was resurfacing going on at the same time. There was, as MOLA geologist Herb Frey likes to put it, a "pre-Noachian" to be discovered; there were stories to be told that predated anything previously thought of. The pervasive layering MOC was discovering, Malin and Edgett told a big meeting on Mars at JPL in 1999, heralded a "new paradigm for Mars geology."

Layers were hardly a complete novelty. Some layering in the walls of Valles Marineris had been evident since the *Mariner* days. There were the layered terrains around the polar caps and some clear evidence of layering elsewhere, most noticeably in what the mappers called the Medusae Fossae formation, which sits on the boundary between highlands and lowlands west of Tharsis and extends, in patches, for thousands of miles. Bits of the Medusae Fossae formation look quite new, other bits look older; some of it looks as though it is in the process of being built up, some of it looks as if it is being eroded away. Layers run parallel in some places, cut across the previous grain in others. Talking to Mike Carr about it, I found him almost charmed by his lack of understanding of the area; both bewildered and amused.

Some argue that the Medusae Fossae formation is an accumulation of windblown sediment from elsewhere on Mars. Others think it might be pyroclastic ash from volcanic activity. Alfred McEwen points out that it is close to the recent flood volcanism that he has mapped in Elysium and Amazonis. The amount of basalt he believes was produced in that flooding was enormous—something similar in volume to the Deccan traps that cover about a third of the surface of India, but spread more thinly and probably produced around ten times quicker. While most of the lava produced would have flowed across the surface smoothly, some would have been pumped into the air in ash-producing "fire fountains"; it would be fair to expect 10 percent of the lava to be turned to ash this way, and since basalt is dense and solid while volcanic ash is light and fluffy the actual volume of ash produced might equal or exceed the amount of basalt. McEwen thinks the thin layers of Medusae Fossae are composed of this fluffy volcanic ash, though others look at the way they have been sculpted and conclude they must be made of sterner stuff.

A more radical theory of Medusae Fossae was proposed in the 1980s by Peter Schultz of Brown University, in Rhode Island. Schultz is a short, sharp, and argumentative scientist quite averse to consensus. On many issues to do with Mars his opinion is a minority one; but not, he will point out quite accurately, a disproven one. Perhaps the most appealing of his ideas has been a Martian application of the phenomenon of true polar wander, something first imagined by the astrophysicist Thomas Gold. If you take a planet with a rigid outer shell—a lithosphere—and add an asymmetric mass to that shell, then under certain conditions the outer shell will slide around in a single piece in order to put that lump on the equator. The planet's axis of rotation will stay the same, since it is only the outer layer that's moving around. But the apparent position of the poles with respect to the rest of the surface will change: They'll seem to wander. On the Earth discussion of such a "true polar wander" fell from favor with the advent of plate tectonics, which showed that though the rigid lithosphere does move with respect to the more fluid mantle beneath, it does so in pieces rather than all at once.* On Mars, though, Schultz thought it might have a role to play.

Looking at Mars it's hard to miss the fact that the lithosphere is indeed carrying a big asymmetric mass, in the form of the Tharsis bulge. Tharsis is almost perfectly centered on the equator, which is either a strange coincidence or evidence that some wandering has been going on. And since on Mars the poles are surrounded by layered terrains, poles that go wandering will leave layers of dust and ice behind them. From these insights Schultz put together a story in which various sets of layered deposits antipodal to each other were seen as fossilized versions of the polar layered terrains. Medusae Fossae was one such fossil. Others were to be found in Arabia, to the south of Elysium, and in the chaoses at the end of Valles Marineris. The path of the poles he derived from these fossils was, he thought, very like the path that would be expected as the lithosphere moved around to accommodate the masses of Tharsis and, to a lesser extent, the volcanoes of Elysium. It's not a story that has convinced many of his peers—some argue he overestimates the possible amount of polar wandering and others think there are better explanations for the layered areas he seeks to explain. It has a wonderful coherence and drama, though. And, like some of Schultz's other work, it makes pervasive layering something that can't just be consigned to a single oddball location like Medusae Fossae. It makes it something that must be looked into all over the high plains.

And also in the canyon depths. Baerbel Lucchitta is a Flagstaff veteran; she arrived on the USGS mesa when her husband Ivo, also a geologist, got a job as part of the Apollo era astrogeology build-up. Having had a baby the day before she was scheduled to defend her Ph.D., she did not immediately join Ivo at work, but not long

*It's because of another form of polar wander associated with plate tectonics that the whole-lithosphere version has to be distinguished as "true" polar wander.

afterward she started taking part-time jobs at the survey that expanded into a full-time career. She mapped the landing site for *Apollo 17;* she trained astronauts; she danced the night away at the Nassau Bay Hilton outside Johnson Space Center. It was, she says, "superexciting." And so, when *Mariner 9* produced images of it, was Mars, which she started studying straight after Apollo and has worked on ever since: Superexciting and much more fun than the moon, even if the only opportunities for dancing (she and Dave Scott had a fine way with the tango) were an occasional evening at the Little America?

One wall of Lucchitta's office is taken up by a vast mosaic of enlarged *Viking* images of the central stretch of Valles Marineris stuck together by a long-gone intern more than a decade ago. It was meant to help her work on a paper about, among other things, the evidence she sees for very recent deposits of volcanic ash in the chasms' depths. The paper was written and published a decade ago, but she's never brought herself to take the vast mosaic down. It's just too fascinating to look at, with its mounds and dunes and fractures. MOC made it more fascinating than ever by revealing that the layering that had been seen by *Viking* only near the tops of the cliffs continued all the way down the canyon walls to the floor. Contrary to prevailing wisdom, the depths were not cut out of undifferentiated basement rock. There was no "primitive" crust, crust that had never been eroded or recycled or squirted out of a volcano; all the way down there were signs of a previous history. And these depths were many miles below the surface of the southern highlands, a surface that had previously been seen as the most ancient thing Mars had to offer.

Looking at the vast image, and imagining its walls layered right down to the floor, it's impossible not to think of the Grand Canyon, fifty miles or so to the north of Lucchitta's office. The comparison between the two is not actually a very helpful one: Olympus Mons is far more analogous to Mauna Loa or Kilimanjaro than Valles Marineris is to the Grand Canyon. The volcanoes differ in size, as do the chasms; but the chasms differ in almost every other way, too. Valles Marineris is straight; the Grand Canyon is curved: Valles Marineris is formed from multiple parallel rifts; the Grand Canyon is a single tear, with tributaries coming in at right angles: Valles Marineris is mostly U-shaped in cross section, with a broad floor; the Grand Canyon is a terraced "V": Valles Marineris does not seem to have been formed by surface erosion; the Grand Canyon obviously was: Valles Marineris contains mysterious mountains added after its creation; the Grand Canyon doesn't.

All this, though, is at some level pedantry. The basic level at which the two are the same is that they are both, in the words of the artist David Hockney, "the world's biggest hole." Though Valles Marineris is a much bigger hole, it is probably not in the end much more striking, if only because the Grand Canyon is already more or less as striking as a view of a hole can be. Hockney, who in the 1990s worked on a series of paintings of it, has a particular love for it precisely because it lies beyond the powers of normal perspective to describe. Representing it forces us to understand something profound about seeing:

> Standing in front of the paintings, your eyes have to move in every direction, just as they do in the Grand Canyon. You have to look in every direction in that great space. Even in the nineteenth century someone observed that there is no perspective in the Grand Canyon. There's no focus point. You simply have to look everywhere: up, down, around. We do that everywhere, really; unfortunately cameras think we always look in one place.

The only previous visual record that comes close to capturing the Grand Canyon is the work of William Holmes, the artist, photographer, and topographer who accompanied the Dutton expedition in the 1870s, and it bears out Hockney's insight. His triptych of drawings from Point Sublime, included in the atlas that accompanied Dutton's monograph, is a masterpiece of geological illustration, recording the strata near and far with an astonishing detail and clarity that no photograph could possibly match. Holmes, like Hockney, abolishes the atmosphere that would otherwise wash out the more distant parts of the vista, representing the huge space as if it were crystal clear. But this is more than a technical illustration; in overcoming fixed perspective to produce a three-piece panorama,

Holmes turned the vast landscape into a superbly executed three-frame comic book. The spaces between the pictures—the turning of the pages of the atlas—transform the images into a story of changing perspectives. In the first frame we look east and see the land pushing in at us in the form of a looming abutment to one of the cloister promontories. In the second the central canyon itself opens up, providing a brief glimpse of the Colorado at its heart; the foreground we're standing on, hardly visible before, stretches out into the image. In the third, the breadth of the canyon as a whole becomes the subject as we look along, rather than across. Over the western cloister the endless riches of the West are revealed.

The astrogeologists at Flagstaff are endlessly aware of the Grand Canyon, and not just because it is what the other USGS people on their mesa study for a living. (It is in that branch of the Survey, as it happens, that Baerbel Lucchitta's husband, Ivo, now works, having transferred there after Apollo.) They go and hike it on weekends; they know it intimately. Such is its appeal that in 1968, at the height of the preparations for the moon landings, Gene Shoemaker took a couple of months off to retrace the steps of John Wesley Powell's pioneering expedition along the Green River and down the Colorado through the Grand Canyon. You couldn't live here as a geologist and not want to do such things; after all, here is a record of five different periods of Earth history, the oldest one and a half billion years earlier than the youngest. There are dozens of rock types, all with their stories to tell, all layered one atop the other; and there are endless lessons about the ways that rock will erode and fall away. You could spend years here without fathoming it all—but each part that you came to understand would be an achievement that added value to the rest. And all of it is beautiful.

By these standards, Martian geology is a rather melancholy thing. Looking at Mars, we are constrained to the single photographic perspective that cameras force on us. We cannot look around; we must focus only on what is in front of us. Though there are ways to milk extra perspectives from the data by using stereographic image pairs and related computerized cleverness, we are for the most part trapped by an orbital viewpoint. We cannot get the sense of movement and narrative that the great space that the Colorado has carved offers those who stand on its rims or descend into its depths—the sense of time not as a dead weight but as a living process. We cannot get that sublime feeling of wanting to jump that comes at the top of such cliffs—the feeling that for reasons that have nothing to do with despair, or even volition, our legs will simply throw us out and over to experience the space in all its openness. (Hockney is quoted as saying that this condition, which affects him as well as me, is called mountain fever. I don't know if the term is used widely but I'm glad to know that other people feel it.)

And we cannot do justice to the details. At a 1999 meeting of the Geological Society of America in Denver, a session on Mars where a number of speakers had failed to turn up was transformed into an impromptu MOC show-and-tell—Ken Edgett had far more slides with him than would fit into his allotted span, so he just Energizer-bunnied along, with the active encouragement of an entranced audience. Image after image flashed by, of dunes and craters and newfound layers. Some were quite like pictures taken by the *Viking* cameras; some were almost unintelligible to the untrained eye. Some simply boggled the mind. My favorite, I think, was of a layered promontory of rock, similar in appearance to the cloisters in Holmes's Grand Canyon panorama but three thousand feet high, or more, jutting into the middle of the crater Terby. Where had those layers come from? And what had carved so much of them away?

Edgett was alive to the excitement the images produced; he wore a little boy's look-what-I've-got-to-show-you expression throughout the presentation. But talking to him later there was also a certain sadness, a frustration that went beyond the eighty-hour weeks with no time to think, NASA's inability to understand how a space mission should be run, the impatience of his colleagues, and all the other things he's been known to complain about from time to time. The problem was that some—most—of this wonderful stuff was simply not going to be understood by anyone for decades, if ever. The MOC archive was already far larger than the *Viking* archive; working through it would be far more work than most people would ever undertake, a substantial portion of someone's life. And even that would not answer the questions. Mars was a big complex planet with lots of surprises. It would take trained humans—people like Edgett and Malin and Schultz and McEwen and both Lucchittas and Bill Hartmann and Dave Scott and Gene Shoemaker and Clarence Dutton and Grove Karl Gilbert—decades of inspired study *in situ* to understand it. The Mars that MOC was revealing was not going to be cracked from orbit. It needed something like the great surveys of the West. It needed years and years of fieldwork. Edgett knows he will never take part in those expeditions or see them completed; he might not even see them begun.

This is what astrogeologists live with. The size and scope of the challenges they face are part of what makes the field exciting. They have a freedom to imagine that is rare in science. But they also have the endless threat of being so goddam wrong—and not even knowing it. It's a humbling, and frustrating, condition of existence. But if it sometimes gets them down, most get back up and just start looking at more pictures. The fact that you can't answer everything doesn't mean you can't answer some things. The fact that you can't see the finish doesn't mean you can't make progress.

Part 3 – Water

—Dear, I know nothing of
Either, but when I try to imagine a faultless love
Or the life to come, what I hear is the murmur
Of underground streams, what I see is a limestone landscape.

—W. H. Auden, "In Praise of Limestone"

Malham

> There is never enough water in my memories of southern Sweden... the brown pine needles that nestled in every crevice of the rocks were dry and sharp as weathered bones. To look at the lakes, and to breathe their damp exhalations, was a kind of healing. I didn't have to catch anything; and not always even to fish. What I needed was to gaze into the surface, and, by gazing, to pass into another world, and breathe.
>
> —Andrew Brown, "Under the Surface"

"Where did your journey to Mars begin?" I asked Mike Carr on a spring afternoon in his office at the USGS campus in Menlo Park, surrounded by a plethora of Martian maps, globes, and photomosaics.

"On the Yorkshire moors," he replied.

Carr grew up in Leeds just after the Second World War and loved to hike in the hills around the city. In much of inland England, especially in the south, the geology is hidden beneath soft soils: In Yorkshire it makes itself apparent. Deep Silurian sediments laid down off the edge of a shattered continent give way first to the sandstones of Devonian deserts and then to limestone reefs from the warm shallow seas of the period that the British call the lower Carboniferous and the Americans call the Mississippian. On top of those sit the millstone grits and coal seams of the upper Carboniferous. That coal fed the furnaces of the great cities of the industrial north. The moors and dales fired their citizens' imaginations.

One of the beauty spots the teenage Carr used to camp at was Malham Cove. It's a concave, overhanging cliff a bit less than three hundred feet high that stands across its valley like a ruined dam, its pale face dotted with rock climbers today just as it was then (though then it was considered one of the hardest climbs in England, while now people seem to see it as relatively routine, whistling the theme from *Mission Impossible* as they pull each other over the last limestone lip). Looking south from the top of the cliff, lower Malhamdale is a gentle idyll of stone-walled fields for sheep and cows, with trees lining the banks of the river that springs from the Earth at the foot of the cove. To the north a strange and savage gully leads through a rock-strewn waste. The limestone beneath your feet, broken and lifted by the Craven fault almost 280 million years ago, has been carved into a weird pavement of slabs separated by cracks a little over an inch, or so, wide—clints—in which sheltered flowers grow. Climb off this pavement and up the gully's far side and you will see one of the most desolate parts of Yorkshire's Pennine Hills. The bulk of Ingleborough skulks under a flat cap of millstone grit in the distance; the still, shallow waters of Malham Tarn reflect the broad, empty sky.

It's not quite the experience of standing on the Kaibab limestone rim of the Grand Canyon, but to see the world from the top of Malham Cove is inspirational enough. It's to be impressed by the wastes to the north, to value the soft fields of the south and to belong to neither. It's to wonder, as Mike Carr wondered, what those hills are made of, and to wonder what shaped them, and to wonder what lies farther off. It's to wonder where the great white slab beneath your feet came from, and how it was lifted up so clean and far. And it's to look down over a cataract almost twice as tall as Niagara's Horseshoe Falls and wonder where the water that cut and polished the rock has gone. In that last respect, Malham is Mars in miniature.

Seeing the rocks of Malham as worked by water, even though the water is not there, is a typical piece of geological explanation: Though there is no water falling now, falling water elsewhere can be seen to cut cliffs in just this way, so this cliff was cut by water. In order to write histories of the rocks they see and the landscapes those rocks form, geologists endlessly have recourse to unwitnessed processes operating in the distant past. One way to make such invocations sound reasonable is to limit the processes available in the past to those observed in the present. This circumscription of possible explanations allowed the "uniformitarian" geologists of the late eighteenth and early nineteenth centuries to differentiate their explanations of the past from those that involved sudden earth-changing catastrophes and to recast their science as one of gradual, observable processes projected back in time. The grandeur of the effects was simply due to the depths of time over which those processes played out.

Thus ancient rocks and landscape structures were to be understood through the study of modern environments that provided analogies to their creation. And the settings for these analogies were by and large watery ones. It is in seas and lakes that you can watch the sediments of today laid down to form the rocks of tomorrow. Rivers and waves carve the valleys and erode the shorelines that will later show up in the strata. Snow feeds the glaciers that grind the Earth. Present water explains past rock, because water is a constant. Layers of rock define geological time; the action of water defines a sort of geological timelessness, an always accessible "now" built into the nature of the science.

If the worlds of the lost past are explained in water, it's hardly surprising. Imaginary waters cover all the greatest discontinuities between worlds. Death lies over the Styx, rebirth on the far shore of Jordan. Afterlives are archipelagos, or gardens where the waters play. Utopias are always islands. In the sixteenth century it was the encircling, isolating ocean that made the New World new, rather than just a continuation of the old. And while sailors were mapping that New World, astronomers turned their nautical telescopes to the moon and could not help but see its two-toned face as one of land and water. The moon was drawn as a set of shorelines, its features defined according to the conventions of maritime charts. When Father Secchi of the Vatican observatory turned his mind to naming the dark and light splotches of Mars two centuries later, naturally he made them continents and seas. What more basic distinction could there be?

The moon's seas had long been seen as illusions by the time that Secchi started his work; Mars's hypothetical waters, protected from evaporation by the impossibility of close inspection, lasted longer. They became integral to the astronomers' understanding of the planet, in that they made sense of various other observations—the polar caps, the seasonal darkenings, the hints of cloud. And if there was life on Mars, as many assumed that there must be, it would naturally require water.

That said, it would not necessarily require much of it. Thanks to Lowell, the twentieth century's myths of Mars were dominated by the scarcity of water. Scarce water was the basis of everything from the despotic power of the Water Worker's Local, Fourth Planet Branch in Philip K. Dick's *Martian Time-Slip* to the religion at the heart of Robert Heinlein's hugely successful *Stranger in a Strange Land*, in which a man from Mars finds the Earth a spiritual desert needing to be revived. The lack of water served to differentiate imagined versions of Mars from the earth—but the fact that there was some water there, just a little, meant that even such an unearthly place could still be related to the human world as an alternative, a warning, or a promise. Even if it was scarce, water still served as a bond between worlds, as well as a barrier. Hence the power of the scene that ends Ray Bradbury's *Martian Chronicles*:

> They reached the canal. It was long and straight and cool and wet and reflective in the night.
>
> "I've always wanted to see a Martian," said Michael. "Where are they, Dad? You promised."
>
> "There they are," said Dad, and he shifted Michael on his shoulder and pointed straight down.
>
> The Martians were there. Timothy began to shiver.
>
> The Martians were there—in the canal—reflected in the water. Timothy and Michael and Robert and Mom and Dad.
>
> The Martians stared back at them for a long, long silent time from the rippling water...

Scarce water could be poetic. But the moonlike pictures provided by *Mariner 4*, combined with the revelation that Mars had barely any atmosphere at all, drove the waters of Mars from scarcity to complete invisibility. There was none on the surface, and the atmosphere contained only a trace of the stuff, enough to cover Mars with a layer just four ten-thousandths of an inch thick. One response to this astonishing aridity was to embrace the idea of a truly unearthlike Mars, dry and deadly. That was what Bruce Murray did, turning the polar caps to carbon dioxide dry ice, reveling in the unexpected dreadfulness of it all, turning the planet's properties into a set of purely physical problems and ruling out the possibility of life.

Then came *Mariner 9,* and the pictures it sent back showed all sorts of features that, if found in Yorkshire, would be ascribed to the action of water. Not marginal features, but a dozen or more huge channels and thousands of little valleys. Yet today the planet offers not a drop. So the science of Mars was forced to take a decisive step away from the uniformitarian canons of terrestrial geology. Its past could not be explained in terms of its present. Indeed, its past might be very different from its present. Though uniformitarian geology had introduced humanity to the shocking idea that time had to be measured in millions of years, it had also reassuringly affirmed that even over such periods nothing really changed. Mountains might rise and fall, but the processes behind those fluctuations remained the same; rain would wear down the peaks, seas would fill with sediment.

Planetary science, though, played out over billions of years, not millions. And from its beginning it had been infected with astronomical modes of thought that were dominated by notions of change over time. Stars, the basic subject matter of astronomy, had been shown to vary greatly in their behavior as they aged. So, surely, must planets. The secret to understanding planets would thus be to understand the general principles of how they evolved over time—principles that would apply, ideally, to all planets in similar situations. Lowell had argued as much himself in his writings on the subject. To Lowell, the differences between the sun's planets could be explained by the effect their differing sizes had on the rates at which they cooled down after their creation. Big planets, with less surface area for their volume, cooled slowly, which was why Jupiter was still hot. Small planets cooled fast, which was why the moon was but a shriveled corpse, "sans air, sans sea, sans life, sans everything." Mars, smaller than the Earth, was distinguished from its sibling by being further along the path to death through desertification that Lowell saw as the end point for all planets.

Mariner 9 seemed to bear out Lowell's view of Mars as being the arid end product of a once more aqueous world. Bruce Murray tried briefly to argue the reverse—that Mars was in fact just coming to life, geologically speaking, a Mundane Egg beginning to hatch—but he didn't get very far. Carl Sagan suggested that Martian history, rather than running downhill, was actually cyclic. Orbital changes, he argued, might cause the Martian climate to oscillate between brief warm "springs," during which liquid water was available if not plentiful, and long cold "winters" during which everything was frozen solid. The Earth, after all, had Ice Ages that appeared to be linked to changes in its orbit (the best evidence for this, which is now seen as conclusive, was published shortly after Sagan's speculation). Why should Mars not have its dry ages?

But crater counting soon made it apparent that the watery features on the face of Mars were much more ancient than such a theory could explain. And James Pollack, a former student of Sagan's at Cornell who had gone on to work at NASA's Ames Research Center south of San Francisco, wove another account of Martian history into an all-embracing review paper that explained not just the surface of Mars but the histories of the Earth and Venus, too—a paper that assembled and extended the work of various researchers into a theory of climate and climate change on the terrestrial planets spanning the entire history of the solar system.

The key to Pollack's paper was the greenhouse effect. The greenhouse effect had been part of Sagan's long-winter model, in which warming due to orbital changes was amplified by the release of water vapor and carbon dioxide from the polar caps. But the caps did not contain nearly enough carbon dioxide for Pollack's purposes. Pollack imagined that early in Martian history the planet had boasted a carbon dioxide atmosphere far thicker than the Earth's atmosphere today. This atmosphere would have been capable of keeping the planet warm enough to sustain an ocean on its surface in its early years—when, according to models of how stars evolve, the sun would have been only 70 percent as bright as it is today—and would thus explain the ancient marks of water on the face of Mars. But it would also have been unstable. It is a basic fact of geochemistry that where you have silicate rocks, water and carbon dioxide, the carbon dioxide will react with the silicates to form carbonates, which will eventually be precipitated out of the water in the form of carbonate sediments—limestones. This would have happened in the Martian oceans and over a period of hundreds of millions of years it would have thinned out the warming atmosphere. Eventually the greenhouse would have failed and the planet would have frozen into the state in which it is seen today.

Pollack's invocation of a thick carbon-dioxide atmosphere was not an ad hoc way of getting the Martian climate warm enough for water. It was a feature planetary scientists had come to suspect was common to all the terrestrial planets in their early history, before differences in size and in distance from the sun had led them to evolve in very different ways. On Venus, too close to the warming sun, the carbon-dioxide greenhouse would fairly quickly have evaporated any ocean, even when the sun was faint. The planet's water would have been lost from the top of its atmosphere as sunlight split its hydrogen from its oxygen; the remaining atmosphere would have got thicker and thicker as volcanoes below continued to spew out carbon dioxide. The end result would be the planet we see today, with a surface hot enough to melt lead and a bone-dry atmosphere a hundred times as thick as the Earth's.

Mars, on the other hand, would have frozen as its carbon dioxide was sucked down into carbonates. And much the same would have happened to the Earth—except for the fact that the Earth was large enough, and thus had enough internal heat, to drive convection currents in its mantle. Convection currents led to plate tectonics, or something similar (the timing of the onset of plate tectonics in the modern sense is not clear). That process would have pushed chunks of oceanic crust down into the mantle, as plate tectonics does, and those ocean floors would frequently have been covered with carbonates. As they sank, the carbonates would have been heated and carbon dioxide given off—carbon dioxide that would have been released back into the atmosphere through volcanoes. Plate tectonics gave the Earth a way of recycling carbon dioxide that on plateless Mars would have been locked away forever. This recycling, Pollack argued, had been vital to life on Earth in the era of the faint, young sun.

The recycling was not perfect—over the aeons, more and more carbonates have managed to stay at the surface—and so the greenhouse effect weakened. Happily, the sun was getting brighter at the same time and so the Earth never froze solid, as Mars did. Recent research suggests it came close a couple of times, with glaciers and sea ice reaching the equator. But when these "snowball earth" conditions prevailed, the ice sealing off the ocean would have stopped the creation of new carbonates. Carbon dioxide recycled through the Earth's volcanoes would thus have been able to build up in the atmosphere, and eventually the greenhouse effect would have become strong enough to melt the ice. On Mars, where there was much less volcanism and no subducted carbonate, such relief was not forthcoming.

Lowell would have loved Pollack's ideas. They were a beautiful example of the comparative methods he had hoped to bequeath to the science he thought of as "planetology." And while Mars's decline into aridity was now to be measured over far greater ages than any canal-building civilization could span, it was still there and it kept its poignancy. A warm, wet Mars might well have brought forth microbial life at roughly the same time that it was emerging on Earth, four billion years ago. That life would not have survived the great drying and cooling, but its fossilized remains might have. So Mars itself, already a ruin, became a fossil, too—the lithified corpse of a living world.

The warm-wet-early-Mars theory also gave a new temporal dimension to the enduring challenge faced by geologists: how to justify the analogies on which their explanations depend. Though the Martian present might be without earthly analogues, the Martian past was not. In its past Mars had rivers and even oceans, and so earthly geological insights could be applied to it. The earthly present, with its waterfalls and rivers, could provide the analogues that would explain the Martian past. Today Mars might be an alien planet with few similarities to our own; in the past, though, it had been relatively earthlike.

One of the main thrusts of Martian geology over the twenty-five years since *Viking* has been the question of deciding which earthly analogues to use when explaining the ancient genesis of Martian features. But the past is not the only thing that has to be explained. From Lowell to Pollack, planetary science had been about change over time. And so explanations of how processes that shape the surface of the Earth now shaped the face of Mars in the past are not, in and of themselves, enough. There have to be explanations of what had happened to Mars since that shaping; of how a more earthlike Mars has become a more alien one. Again, water lies between the worlds as bond and barrier. There have to be explanations of where the waters of the early, more earthlike Mars have gone to, of how much water there ever was, and of how much might be left.

Mike Carr's Mars

> Forms so new to the culture of civilized races and so strongly contrasted with those that have been the ideals of thirty generations of white men cannot indeed be appreciated after the study of a single hour or day.
>
> —Clarence Dutton, *Tertiary History of the Grand Canyon District*

Mike Carr was to spend more than half of his professional life putting together the story of water on Mars. How he came to do so was largely a matter of chance. The interest in rocks and landscape he'd developed on the Yorkshire moors sent him to the University of London to read geology. Poor employment prospects—the end of empire meant that Britain's geologists had first call on a much-reduced allotment of the planet's geology—sent him to Yale for graduate work. A distaste for working in the field during the American summer—too hot for a man used to a cooler, grayer world—led him to concentrate on lab-based geochemistry. When his student days, and student visa, came to an end he moved sideways again, taking up geophysical research on shock waves at the University of Western Ontario, research that was part of a military program looking at the effects of nuclear explosions on underground facilities. Reading the paper by Gene Shoemaker and Edward Chao on the formation of coesite at impact craters in 1961—and having met Danny Milton, one of the first USGS astrogeologists, at a Yale reunion—he wrote to Shoemaker about his own work on inducing similar changes in rocks using high-pressure techniques in the lab. A matter of days later Gene was on his doorstep demanding that he come to work at the Survey. He did, moving to Menlo Park late in 1962 and not leaving since.

He did not get to Mars until a bit later. Like almost everyone in the astrogeology division, Carr was assigned quadrangles of the moon for geological mapping—it was more or less a condition of employment, cold nights at the Lick Observatory above Santa Cruz picking out details the cameras couldn't catch a rite of passage. He carried on with his experiments on the effects of shock waves until a laboratory explosion tore off parts of a couple of fingers on his left hand, at which point, reasonably enough, he stopped. He also got involved in a tedious and classified military program on the effects of radioactive dust. By 1970, though, he was on his way to Mars. He was one of the original USGS members of Masursky's *Mariner 9* television team and was put in charge of the more sophisticated cameras on the two *Viking* orbiters. In the 1970s, no one saw more of Mars than Mike Carr did.

Carr's early Mars work, coming from his time on the *Mariner 9* team, was on the planet's most obvious features, the volcanoes. But the *Viking* images, far more detailed and numerous than those from *Mariner 9*, seduced him away from the largest features and toward the sheer variety of landscapes the planet had to offer. He used the *Viking* pictures as the basis of a book, *The Surface of Mars*, which is still unrivaled as a single-author survey of the subject. And he used it to probe the mystery of the missing water.

In 1975 Bob Sharp (the man who had brought Bruce Murray on to the *Mariner* team and supervised Gene Shoemaker's master's dissertation) and Mike Malin, then also at Caltech, divided the Martian features that appeared to have been caused by fluid erosion into three distinct categories: outflow channels, fretted channels, and run-off channels. The outflow channels were the most dramatic; many hundreds of miles long, often tens of miles across and sometimes wider. These channels typically started from the strange, chaotic terrains that Sharp had first observed in the *Mariner 6* and *7* data, areas where the surface seemed to have been undermined and then collapsed over thousands of square miles. Downstream they often spread out into a number of subchannels that diverged and recombined like a wide stream running over coarse sand—a process known as anastomosis—before eventually fading away. Within the anastomotic areas were beautiful streamlined islands; hardened crater walls created vast teardrops in the stream.

Some of these channels emptied into the Amazonis plains on the west side of Tharsis; a couple more ran down into Hellas, and a couple more flowed from the flanks of the Elysium volcanic massif. But these, for the most part, were the smaller ones. It was the channels draining into Chryse, north of Valles Marineris, that were truly extraordinary, with calculations based on their width and slope suggesting they were carved in days or weeks by floods of as much as half a cubic kilometer of water a second—ten thousand times the flow of the Mississippi. Imagine all the great lakes emptied into the Gulf of Mexico in just a couple of weeks and you get some idea of the scale.

Smaller than the outflow channels, but still impressively large, were what Sharp and Malin called the fretted channels. These were mostly found along the northernmost reach of the great escarpment between the highlands and the lowlands, and unlike the outflow channels they had very steep walls and flat floors six miles or more across. Like the outflow channels, and unlike most earthly valleys, they often didn't have tributaries. Indeed, there was no clear evidence that anything had ever flowed along them at all. They looked like cracks that had somehow eaten back into the highland material, widening as they did so. Of all the channels on Mars, those in the fretted terrains looked least earthlike.

The most obviously earthlike were the valleys that Sharp and Malin called run-off channels. Run-off channels had tributaries that fed into them; they started small and got wider and deeper downstream. In short, they looked like terrestrial river valleys.* These valley networks looked earthlike and they also appeared to be pretty common. A graduate student named David Pieri counted two thousand of them on the *Mariner 9* images; when the *Viking* images started to come in they showed yet more. Valley networks hundreds of miles long were spread across much of the southern

*Indeed, to most people they seemed so like valleys that the term "channel" was deemed a mistake; to geographers and geologists, channels are features that are at least sometimes filled by flowing liquids. A river's channel is the space between its banks. A river's valley may be a far wider trough that it has cut into the landscape over time, or a physical flaw in the Earth within which it happens to lie. Martian geologists now prefer the term "valley network" to Sharp's and Malin's "run-off channel."

highlands. Far more than the catastrophic floods and the strange frets, these valley networks evoked an earthlike planet, one where it was warm enough for rain to fall from the sky and flow across the surface.

In a series of papers in the late 1970s and 1980s—some written while he had Shoemaker's old job as head of the USGS astrogeology branch—Carr looked at the various channels and valleys, and became convinced that Mars must still be relatively well endowed with now hidden water. Various lines of evidence—the shapes of craters, the slumping of landforms, the strange cracks in some parts of the surface—led him to believe that much of that water was now sitting in the high latitudes as dirty ground ice. Most geologists looking at the valleys agreed with Carr that the planet had a fair bit of water. But geologists were not the only people studying the planet.

Many of the geochemists looking at the planet's chemical make-up thought Mars should have very little water. One set of geochemical arguments, based on the surprisingly small level of argon found in the Martian atmosphere by the *Viking* landers, suggested that there had never been much water there in the first place. Another set, based on the amount of deuterium—heavy hydrogen—in the small amount of water vapor in the atmosphere, suggested that most of the planet's water had been lost to space. After *Viking* it was agreed that there was water ice in the north polar ice cap, where the summers are cooler than they are in the south. But there was not that much of it, and at the landing sites the aridity was extreme. These findings and the geochemical arguments made some nongeologists skeptical of water-based explanations for the channels and convinced most of the others that if there had been water in the atmosphere during an early warm, wet phase of the planet's history it had long since been lost to space. Carr, perhaps emboldened by journeyman years that had grounded him in geochemistry and geophysics, pointed to inconsistencies in some of the geochemical arguments and at alternative interpretations for others. But he laid the most weight on the evidence from the ground—evidence that seemed to argue unequivocally for the presence of water in the past and ice in the present.

In this, Carr was echoing the nineteenth-century geologists who insisted that the Earth was far more ancient than physics would allow, or the twentieth-century palaeontologists who insisted that the fossilized fauna of South America and of Africa were so similar that the two continents must once have been one—something that geophysicists would not hear of. Geological lore is replete with attempts by outsiders armed with calculations to tell geologists that what seems clearly written in the rocks themselves cannot in fact have happened. The fact that geology is a science of interpretation, rather than deduction, makes it peculiarly vulnerable to being trumped by seemingly more rigorous disciplines in this way. Uniformitarianism is one reaction to such situations—if you only ever evoke forces already visible in the world around you, few will tell you your explanation cannot be true.

The other reaction is simply to stand firm—to rebuff the hegemons of higher theory with Columbia geology professor Marshall Kay's robust restatement of Kant's view that the actual proves the possible: "Anything that *has* happened, *can*." If you are going to take this approach, though, you need to make your account of what has happened as clear and compelling as possible. Being able to express it in mathematical terms will probably help, too—chemists and physicists like that sort of thing. And that was exactly the sort of thing that Carr gave them, measuring his ideas down to the last sverdrup (a measure of water flow named after a Norwegian oceanographer) and darcy (a measure of the permeability of rock named after the French hydrologist who built the public fountains of Dijon). He ended up estimating that the amount of water on Mars was enough to cover the surface globally to a depth of around four hundred meters (1,300 feet). That makes Mars a dry place compared with the Earth, where the equivalent figure is a couple of miles, but it makes it a very wet place compared with the figures that geochemists were throwing around of a few meters.

Like every geologist who has made a lasting contribution to the study of Mars, Carr is a patient man. (He also has a tendency to plow his own furrow. Though he's done more than his share of committee work, many of his papers have no co-author, which is rare in the field. When Larry Soderblom took over as head of the Survey's astrogeology branch in the 1980s, he's said to have teased his team by telling them he would combine the financial acumen of Hal Masursky with the people skills of Mike Carr.) Carr was quite willing to sit down and look at *Viking* images day after day, studying each one in detail. During the late 1970s he and a colleague went through seven thousand of them, picking out channels and valleys, and marking their courses on a set of the USGS's 1:5,000,000 maps. These marked-up maps were then digitized and compared with the geological map of the planet that Carr had worked on with Dave Scott. The comparison produced strong statistical support for the already widespread understanding that the valley networks, the best evidence for persistent running water, were restricted almost entirely to the very oldest Martian terrains. While it was possible that this had something to do with the rocks that made up those terrains being more easily eroded, the obvious conclusion was that running water had been a feature of only the very earliest days of the planet's history. Almost as soon as the heavy bombardment of those days was over, the streams dried up—and the water went underground.

A boyhood walking over limestone moorland riddled with sinkholes and potholes had taught Carr that planets need not be as solid as they look, and he found various reasons to imagine that the Martian crust was pretty porous. The Apollo missions' seismometers had revealed that the top fifteen miles of the moon were apparently made of chunks of cracked and faulted rubble called "megaregolith," a result of the fact that the moon has been hit with large rocks on a regular basis. Since Mars was made of similar rocks, and got hit just as much, Carr reckoned it should have a similar megaregolith full of accommodating pore spaces. Because gravity was stronger on Mars than on the moon the Martian megaregolith might not go so deep. But just six miles of porous rock can soak up a lot of water. In some places you could imagine you were seeing the water sink straight into this underworld. The main trunk of one of the longer and more prominent valleys in the south, Nirgal Vallis, seems at one point to disappear for tens of miles before resurfacing to continue its course—just like the stream that vanishes into the ground above Malham Cove and then emerges from the rock at its foot.

Carr imagined that much of the planet's water had sunk into these subterranean aquifers early on. Heat from the planet's core would have kept the water liquid—except near the surface, where it would freeze. Frozen water in the uppermost crust would then act as an impermeable barrier isolating the rest of the water from surface conditions. As time went on and the planet cooled, that barrier would grow thicker. And if Mars had been flat, or the rock in the aquifers relatively impermeable, that would have been the end of the story: The water would have been interred for the duration.

But the rocks of Mars are largely basalt, which can be pretty permeable. And Mars is far from flat. Imagine a large aquifer system spanning the borderland between highlands and lowlands. In the lowlands, the water in the aquifer would be pressed up against the bottom of the layer of frozen ground immediately below the surface. In the highland part of the aquifer, though, the water level might be as much as three miles higher. The weight of that water perched in the highlands would impose a huge pressure on the lowland aquifer and its permafrost cap. If that cap were broken by an asteroid impact or a Marsquake, a mountain range of water would tumble out of the highlands and onto the surface of the lowlands. Near the breakout point the surface would collapse as the highly pressurized water it had been resting on spewed forth like rocket exhaust; downstream a thousand-Amazon flood would tear across the landscape in an orgy of abrasion, creating one of Mars's vast outflow channels.

People had suggested that the floodwaters came from underground before, but no one had thought the possibility through as thoroughly as Carr had. With the help of colleagues elsewhere in the USGS, he'd taught himself a fair amount of hydrology and the numbers in his account added up convincingly. One of its great strengths was that it explained how huge floods could come from relatively small chaotic terrains. In Carr's theory the chaos areas were not the sources; they simply marked sites where the worst of the subsurface erosion associated with the emptying of the aquifer had caused the ground above to collapse. The same reasoning explained why there were some chaotic terrains uphill of the sources of outflow channels but not directly connected to them.

The theory also accounted for the timing and location of the channels, at least in a hand-waving sort of way. The channels formed well after the valley networks in the southern highlands, and at more or less the same time, according to some theories, as the rise of Tharsis. The rise of Tharsis led to a downward flexure of the crust elsewhere, notably in Chryse. That would have exacerbated the local pressure differential between the highlands and the lowlands, leading to breakouts in the lowest parts of the new-formed trough.

The aquifer argument also threw light on some of the oddities of the canyons of Valles Marineris. The canyons, too, appeared to have formed at least in part in response to the rise of Tharsis; the crust had stretched and as a result some blocks had settled to a much lower level than the surrounding surface (though subsurface erosion may have played a role here too). This lowering of the canyon floors would have meant that, in some places, the surface would have ended up below the aquifers' water level. Water would have flowed out of the aquifers and into the canyons, where it would have formed large, deep lakes, their surfaces frozen but their depths still liquid. As Jack McCauley pointed out in the 1970s, sedimentation in such lakes might explain the stacks of layered material that rise up from some of the canyon floors. When the permafrost layers in the lowlands to the north and east of the canyon system ruptured, the lakes would have drained away through the aquifers' secret passages, leaving table-mountains of sediment behind.

Where did those floodwaters end up? They would have flowed downslope to the north, rafts and sheets of ice constantly forming at their surfaces, until they pooled in the lowland plains and sank into their sands. Most of the water would end up as permafrost; some remnant lakes would freeze into lenses of solid ice, slowly to be covered with windblown dust and sand. Such ice deposits, covered by a veneer of windblown sediment, might make up much of the lowlands.

The story that Carr pieced together from his own work and that of others did a good job of explaining many of the watery features of Mars. It also explained why there were not more of them. The idea that the water was sequestered underground or in stable, high-latitude ice deposits explained why there was relatively little erosion on Mars. If Mars had been warm and wet for any significant fraction of the time since the end of the heavy bombardment recorded in the Noachian cratered highlands, those craters of the highlands would be heavily eroded; they aren't. The idea that most of the water had been underground almost all of that time explained the lack of erosion nicely.

It also helped explain the high deuterium levels that seemed to indicate that Mars's water had escaped from the planet altogether. Water molecules in the upper atmosphere of Mars (or the Earth, or Venus—this is another of those properties that are the same in all three places) risk being destroyed by ultraviolet rays from the sun; if the hydrogen atom let loose in this process then escapes from the planet altogether, the shattered water molecule can never be reconstituted and water is lost for good. Since deuterium is heavier than hydrogen, it is less readily lost than hydrogen through this process, so as time goes by, the ratio of deuterium to hydrogen in the water left behind will rise. On the Earth, almost all the water in the atmosphere is trapped in the lowest levels, so not much is lost. In the supergreenhouse atmosphere of Venus, into which an ocean's worth of water presumably evaporated early in the solar system's history, water used to get right to the top—so almost all of it has been lost. As a result, the deuterium ratio on Venus is a hundred times what it is on Earth. In the Martian atmosphere the ratio is about five times what it is on Earth. If the ratios started out the same, this level of enrichment suggested that as much as 99 percent of Mars's water has been lost (probably early in the planet's history). But that would be true only if the water on Mars had regularly been cycled through the atmosphere, as it is on Earth (and was on Venus, before it all got removed). On Carr's Mars, most of the water spent most of its time locked away in the subsurface; the water currently in the atmosphere would represent only a little more than a billionth of the planet's total reserves. Most of those reserves, Carr predicted, will turn out to show far less deuterium enrichment.

By the 1990s Carr had convinced most of those who were in need of convincing that there was a lot of water frozen into the Martian crust. But there were still many details to be sorted out, most notably those concerning the earliest watery features, the valley networks. On Earth, especially in temperate climates, valley networks tend to saturate land surfaces; there's always a tributary to something nearby. If the Martian valley networks had drained water from the surface of the ancient southern plains in the planet's warm, wet youth, their tributaries should have covered the plains. But they didn't; the networks mostly had only a few tributaries and were often far apart from one another, leaving large parts of the plains undissected.

One way of explaining this was to suggest that the water that cut the valleys did so from below, in a process known as "sapping." Sapping starts when a dip in the surface lets out water from an underground aquifer. The flow of water undermines the rocks and soil above and behind the spring. As the undermined surface collapses the spring eats its way farther and farther into the higher ground—a process called headward erosion—and leaves a valley trailing behind it. You can see such things throughout the arid parts of the American West; Arizona is rife with them. In some ways the Martian valley networks looked quite a lot like sapping channels; they often had V-shaped cross sections and tended to originate from rounded half-bowls.

One advantage of sapping was that it didn't need rainfall. In the early 1990s the idea of a Mars that had truly been warm and wet in its early history came in for some criticism. According to new climate models no amount of carbon dioxide could produce the truly phenomenal greenhousing required for open water and a maritime climate. Mars had certainly been warmer and wetter, the critics said—but seeing as it was now spectacularly cold and dry, "warmer and wetter" could still mean a lot colder and drier than the Earth. Sapping, with its water coming out from within, had the attraction of explaining the valley networks without recourse to rain and thus not requiring a planet warm enough for a moist atmosphere.

Carr saw that this attraction was partial, at best; sapping might not require rainfall, but it did require that the aquifers feeding the process be replenished, and he saw no convincing way of doing that over a wide area other than by having water fall somewhere on the surface and trickle down through the permeable rocks. Instead, in the 1990s, he developed a theory that made the valley networks through a process that might not have required any rain at all. Rather than seeing the valleys as sapping structures, he turned them into long landslides helped along by lubricating ice or groundwater. Such a "mass wasting" mechanism would explain why, unlike river valleys, the Martian valley networks mostly showed no sign of fluid-carrying channels winding along their floors. It would explain why they sometimes reached right to the top of crater walls, where there could not possibly be any aquifer to feed them. It would make them much closer relations to the strange fretted channels found at high latitudes, also attributed to some sort of ice-associated mass wasting. And it would allow them to be created with less water than would be needed for run-off or sapping.

This model did not convince his colleagues; it doesn't wholly convince Carr these days either, largely because the image torrent from MOC has shown that the valley networks are highly varied. Some may be built by mass wasting, some by sapping, some by small floods. A few may even have formed by surface run-off, though for the most part the fine-detailed images showed very little evidence of the small tributaries surface flow would produce. But if the mass-wasting model wasn't a complete success, it did encapsulate the rather idiosyncratic attitude that Carr has developed toward the waters of Mars. In some ways he is their champion. It was he who consolidated the case for their existence in floods and ground ice and aquifers; it was he who fought for a full four hundred meters' worth in the face of geochemical scepticism. And yet if there is a more watery and a less watery way to get some landscape feature made, Carr will always plump for the less watery one unless it involves him in clear contradictions. There is something profoundly parsimonious in his attitude, as though every extra thousand cubic kilometers of water or degree of surface temperature that he allows the planet is subtracted from a store elsewhere for which he will be held accountable. When people argue for the planet having more water than he willingly allows—as many have done in the past decade—Carr will dig in his heels at their profligacy. And Carr with his heels dug in takes a lot of moving. He's tenacious, even stubborn. Stubborn as a rock.

Reflections

—I love all waste
And solitary places; where we taste
The pleasure of believing what we see
Is boundless, as we wish our souls to be:
And such was this wide ocean, and this shore
More barren than its billows;—

—Shelley, "Julian and Maddalo"

Carr's *The Surface of Mars* was to have a great influence. For a decade, it was the only major scientific text that dealt with every aspect of the planet's geology in the light of the results from the *Viking* missions. As such, it interested not just scientists, but also science fiction writers—most notably Kim Stanley Robinson. After some success with his first novel, *The Wild Shore,* written while he was studying literature at the University of California, San Diego, Robinson thought it might be worth revising an earlier unsold novel, *The Grand Tour,* which dealt with a trip through the solar system from Pluto to Mercury. Revising the segment on Mars, he turned to Carr's book and, he says, "just realized what a spectacular planet it was." The idea of a novel that would do justice to the spectacle of Mars became an ambitious new goal. It would be an epic story of transformation, of water where there had been none before, of a planet new born. It would be called *Green Mars.*

At the time Robinson was being inspired by Carr's well-illustrated and precisely argued pages, science fiction's interest in Mars was at an historic low. Imaginations used to a Mars that was merely arid had shriveled in the face of the moonlike waste and utter desiccation revealed by the first *Mariners*. In the mid-1970s, at the time of the *Viking* missions, only two major pieces of science fiction dealt with the planet in a remotely realistic way, *Man Plus* by the American veteran Fred Pohl and *The Martian Inca* by British newcomer Ian Watson. Both are bleak. In *The Martian Inca,* Martian life weathers the millennia-long winters between Sagan's springs in the form of self-organizing spores, and these turn out to have bizarre effects on human consciousness; on Mars these effects lead to the first human landing failing, while on Earth they lie behind a brief, bloody, and futile attempt to revive Incan theocracy on the Bolivian *altiplano.* In *Man Plus* the hostility of the Martian surface requires that an astronaut going there be reengineered as a cyborg if he is to survive. His human contacts, physiology, senses, and even genitalia are sacrificed in the technological makeover. Both books transmute the profoundly alien nature of Mars into the alienation required of those who touch it or are touched by it. Mars is a world that can only be reached if humanity is left behind. The journey between planets is as much like a death as it is a sea voyage to a distant land.

Both those novels take the science of Mars seriously—the lack of water and air, the temperature, the dust storms. Watson still remembers the day in 1976 when the bound proofs of *The Martian Inca* arrived through the letter box of his home for urgent correction; in the same post came the latest issue of *New Scientist* magazine with a picture of Mars on the cover and a headline screaming "Water Ice on Mars" (a reference to the fact that the *Viking* orbiters had provided evidence that the permanent northern polar cap was made of frozen water rather than carbon dioxide). To his dismay it seemed that various details of *The Martian Inca* were out of date before it was even published. Within an hour, though, fate had intervened in the form of Greg Benford, an American scientist and science fiction writer visiting the Institute for Astronomy in Cambridge. Benford arrived at Watson's house unannounced, full of news about *Viking* (he had visited JPL only days before), and had his brains thoroughly picked. The next day Watson rewrote phrases, sentences, and even paragraphs of the book to reflect the Mars uncovered by *Viking,* all without making any given page longer or shorter—a constraint imposed by the fact that the page numbers were already set. This act of fealty to the real was, he says, "a feat of which I remain mildly proud."

But if Pohl and Watson took the science seriously, their concern was more with broad concepts than with specific places; both books offer only a single landing site as scenery, in Pohl's case not even saying where on the planet it is. (Watson's astronauts land where Mangala Vallis, an outflow channel, opens out onto Amazonis Planitia; this was, as it happens, one of Hal Masursky's favorite candidate landing sites.) When Robinson read Carr, though, it was the range of landscapes and their specificity, their already known certainties that entranced him. He saw that it was, as he later put it in an interview with David Seed for the journal *Foundation*, "A most amazing opportunity for a science fiction writer—a whole world given to us that is right next door, and real, but empty."

The Mars that the *Mariners* and *Vikings* had revealed offered Robinson a way to reconcile two of his driving concerns—an interest in writing science fiction that meets the critical standards of mainstream literature and a deep commitment to environmentalism. The concerns are linked by a sensitivity to setting. Most science-fictional worlds are relatively thin places, stage sets in front of which a drama suggested by some scientific possibility will unfold. When more depth is attempted, the thickening world that ensues—unattached to the reader's experience of reality—quite quickly begins to feel more like the inside of the author's head than a place to which one might actually travel. Good fiction is quite possible under these conditions, but the naturalism of a certain type of closely observed contemporary novel is not. Robinson responds to and aspires to that sort of naturalism, and he realized that a mapped Mars offered a new way of achieving it—one in which contemporary science provided not just ideas, but also the details of the setting. Science fiction had abandoned Mars because science had proclaimed the planet lifeless, and lifelessness was not something science fiction knew how to articulate. But the same science had brought into being an extraordinarily documented, well-mapped planet, a setting whose coherence could be guaranteed by something more objective than whatever discipline the author might bring to imagining it.

That sort of respect for the world fits nicely with an environmentalist viewpoint. Robinson's commitment to the environment is evident in what he writes, in what he says to the science fiction fans who invite him to their conventions, and in where he lives. He and his wife Lisa Nowell (who works for the USGS) live with their children in a lovely development on the outskirts of Davis, California, called Village Homes. It's a place where the houses face bike paths, not roads; where water management is by means of natural ponding, not concrete gutters; where land is cultivated in small parcels, both individually and communally. Robinson grows some lovely, tiny alpine strawberries, among other things. It is perhaps unfashionably hippyish—all the street names are taken from Tolkien—but it is also sincerely and pragmatically utopian, a green Californian attempt to rethink suburbia and do it properly. A very similar impulse dominates Robinson's Mars books; they are a green Californian attempt to rethink humanity's relationship with the planet it lives on and to do it properly. He imagines a Mars where humans engineer massive climate change, increasing the temperature by thickening the atmosphere, releasing water and gas frozen into the soil, and generally making the place more habitable.

On such a "terraformed" world, the living environment would be an utterly central concern, the growing civilization's greatest creation and its greatest responsibility. And leaving Mars to one side, that's just the attitude to the environment that Robinson would advocate on Earth. This concern made Robinson's fidelity to the Martian landscape a matter of moral seriousness as well as aesthetics. It also offered what science fiction had been missing—a way to deal with the lifelessness of Mars that did not alienate the reader. Robinson had no need of indigenous Martians to fulfill the old fictional role of offering the insights of the "Other" to the visiting humans. For Robinson, the role of the Other was played by the basic raw material of the planet, material that would shape the protagonists' world.

Newly excited by Mars in the process of revising *The Grand Tour,** and knowing that it would be a while before he would write the Great Martian Novel—*The Wild Shore* had given rise to two further novels linked to it, one of which would be the Californian utopia *Pacific Edge*—Robinson wrote a novella, *Green Mars,* as a sort of promissory note. It established a moral right over the title

*Eventually published as *The Memory of Whiteness.*

and it explored what Robinson knew would be the central issue in his eventual novel—the changes that humanity would inflict on the Martian environment in the name of terraforming and the resistance that some people would feel toward such change. Only in 1988 did he start to write the novel itself, and it soon became apparent that his vision was too broad to fit between one set of covers. It eventually appeared in three volumes—*Red Mars*, *Green Mars*, and *Blue Mars*—and took up a decade of his life. The books'* seventeen hundred pages take the story of humanity on Mars from the early twenty-first century to the twenty-third, and from the Mars described by Carr to a Mars wholly Robinson's own. They are the tale of a planet becoming a world.

Two long slow stories run through the books, that of political independence and that of terraforming. Neither is new. Almost every American science fiction novel about a Martian colony involves some sort of rebellion—at some level, the stardate is always 1776. Robinson's revolution, though, is not a typical product of the genre. For a start, it unfolds in stages over centuries, with the first attempt to shuck the surly bonds of Earth spectacularly unsuccessful. And when it starts to come right, it is a revolution driven by something grander than a wish to throw out the Earth's redcoats. It is a revolution against corporate capital, to some extent a revolution against all property not communal. It is an attempt to set up a utopia that might have been recognizable to Marx, or to Kropotkin. While this is unusual in American science fiction, a Mars red in more ways than one is not without precedent in the planet's lore. A century ago the Martians' assumed antiquity was sometimes seen as proving that their civilization must be further evolved than ours, superior, peaceful, and quite possibly socialist. Lowell's canals, a planet-wide technical achievement in the face of disastrous drought, were seen as having become the moral equivalent of war and of having ushered in an age of peace and harmonious productivity.

*Robinson doesn't like to call the three books a trilogy, largely because trilogies in science fiction have a reputation for being a purely mercenary way of milking the popularity of an initial story or set of ideas; he sees the work as a single three-volume novel. I, for what it's worth, see it as two linked novels, with *Red Mars* a freestanding beginning and *Green Mars* and *Blue Mars* the two halves of its sequel.

Edward Henry Clements, editor of the *Boston Transcript,* devoted four hundred lines of iambic pentameter to celebrating this idea, looking forward to the day when canal-building engineers would replace imperialist butchers as the heroes of the Earth. "In short," he wrote to Lowell, "I am going to show why Mars is carrying through our Heavens the heart-red flag of socialism." The Russian writer Alexander Malinovsky wrote a sequence of works about a socialist Mars before and after the Russian Revolution, culminating in a 1924 poem called "The Red Star: The First Bolshevik Utopia." One of Robinson's characters is named for Malinovsky's pseudonym, Alexander Bogdanov.

Like revolution, the idea of terraforming—a term invented by the science fiction author Jack Williamson in the 1940s—is also well rehearsed within the genre. One of science fiction's enduring successes, Frank Herbert's *Dune,* deals implicitly with the terraforming of a fictional planet, Arrakis, clearly modeled on Mars* (the terraforming becomes more explicit in the many sequels). The doomed mission in *The Martian Inca* was an attempt at terraforming; it was precisely because the engineering of environments to suit humans was such a commonplace in science fiction that Pohl's evocation of the opposite process in *Man Plus* created a stir within the field. Jerry Pournelle's novel *Birth of Fire,* the first novel to make full use of the dramatic possibilities revealed by *Mariner 9,* conflates terraforming with revolutionary war. But if the subject matter of his stories was not new, Robinson gave it a new literary richness, historic sweep—and topographic detail.

He was not the first writer to use maps of Mars to add to his fiction. Robert Heinlein's *Red Planet* borrows its geography from Lowell's charts, Arthur Clarke's *The Sand of Mars* (the first Martian terraforming novel) uses Antoniadi. Edgar Rice Burroughs may have based Barsoom—his Martians' name for Mars—on a number of maps, though the links between his landscapes and the then

*"Observe closely, Piter, and you, too, Feyd-Rautha, my darling," says the evil Baron Harkonnen near the beginning of *Dune,* as he gestures at a globe. "From sixty degrees north to seventy degrees south—these exquisite ripples. Their coloring: does it not remind you of sweet caramels? And nowhere do you see blue of lakes or rivers or seas. And these lovely polar caps—so small. Could anyone mistake this place?" No.

mapped Mars remain fairly obscure. But Robinson had the wonderful USGS maps drawn by Bridges and Inge, and he intended to make full use of them. Where most Mars fiction, like Pohl's and Watson's, concentrates on a single setting, Robinson's characters travel Mars from pole to pole, from mountain peak to canyon floor. They crawl under its ice caps and down its canyons, they set up laboratories inside its great cliffs and build cities on its mountains. Looking at the maps on my wall, I can see few places where some incident in the books does not take place. Sitting in Robinson's home office in Davis, it was a delight to pore over the maps and globes on which he marked out journeys, towns, and shorelines as he was creating them. Robinson loves maps of all sorts. When he lived in Switzerland he got hold of a huge embossed plastic relief map of the Alps with which to plan his hikes, scrutinizing possible panoramas in advance. Armed with the airbrush artistry of Bridges and Inge, he produced a profoundly topographic fiction, one in which the shape of the planet and the appearance of its features repeatedly allow a reality independent of the author to impress itself onto the characters.

Robinson's books contain the most textured and varied evocations of a mapped Mars that literature has to offer. That's not to say they are perfect. While in many ways Robinson is wonderfully true to the unearthly scale of Mars—early on in *Red Mars,* an immigrant is derided for seeing the quasi-continental bulk of Olympus Mons as just another shield volcano—he can't resist assembling the features on the maps into landscapes for his characters to experience even when the scale is too vast to permit it. For example, if you look at a USGS map of the Tharsis plateau it's hard to believe that standing on that high and otherwise largely featureless plain one would not see all its vast volcanoes. They look like a family, like they should know each other—like Mounts Rainier and St. Helens and Adams and Hood in the Cascades. But the spherical geometry of a planetary surface and the opacity of a real atmosphere breaks the family apart. When Robinson's characters drive between Ascraeus Mons and Olympus Mons they see the mountains looming to either side of them. But Ascraeus and Olympus are separated by a twelfth of the circumference of the planet. In geometrical terms they're roughly as far apart as San Diego, California, and Savannah, Georgia, or as Luanda on Angola's Atlantic coast and Dar es Salaam on the shores of the Indian Ocean. Sunset at Ascraeus is midafternoon on Olympus; a thousand miles of dusty air lies between them. The maps suggest more than the eye could ever see; no one vista can ever give you a continent of truth.

The maps gave Robinson more than just the lay of the land; they also helped give him its language. The books relish the Schiaparellian names already given to features on Mars and the sense of the ancient they can evoke. The wonderfully euphonious names of the valleys—Kasei, Tiu, Nirgal, Ma'adim—become the names the first settlers give to their children, and come to form a chant of acclamation by which the colonists celebrate their planet under all its ancient names.* New names have their own logic. The first settlement is called Underhill, which is both descriptive and a reference to the promise of renewal that lies at the heart of Arthurian legend. Some towns and cities are named for places on earth—Cairo, Nicosia, Odessa—some for people. Edgar Rice Burroughs gets the capital on the slopes of Isidis basin, fellow science fiction writer Charles Sheffield—who helped with some of the details that Robinson needed—gets a city on top of Pavonis Mons (for reasons that will become clear later in this book). There's a town called Carr, as well. The protagonists' names are chosen with precision and relish: The woman most committed to keeping the surface of Mars as it is today, characterized by her deep affinity for that which is mineral, is Clayborne; the chief scientist who brings life to the surface is named Saxifrage, for the alpine flower whose Latin root means breaker of rocks; the Russian utopian is Arkady. The great underground aquifers from which the water flows (aquifers that Robinson keeps far too well stocked for Mike Carr's tastes) are all named for the writers of earlier Martian fiction. As the Martian settlers release the waters to the surface, so Mars moves closer to the buried dreams of those early writers and their imagined earthlike world.

*Gary Snyder, whose poetry and whose love of the wilderness are strong influences on Robinson, did something similar in "Turtle Island" with words for "bear" in different languages.

As that world changes, so do the sources of Robinson's realism. As its inhabitants slowly build something utopian, if not a utopia, Mars comes alive with delicate beauty, with landscapes drawn not just from maps but from experience. Most environmentalists—and perhaps most people—have a landscape they relish above all, an ensemble of sights and sounds and smells through which the Earth speaks to them most clearly. In Robinson's case the sacred landscape is that of the recently glaciated fell-fields in California's High Sierras. Once a year he and his oldest friends head up to this high country for a week or so, and it's easy to get the impression that those weeks are the most important constant in his life other than the central concerns of his family and his artistry. The mixture of strength and subtlety in the sparse landscape between the tree line and the snows clearly entrances him; when a version of that same postglacial landscape first makes its appearance on Robinson's Mars, about a third of the way into the second volume, his delight lifts the description from the page. While many other living landscapes grace the books from then on, it is always this alpine biota, at once so enduring and so seemingly fragile, that lives in the imagination most fully, establishing a new bridge between the books' personal and planetary scales.

There's a seeming contradiction here. The surface of Mars is extraordinarily ancient; the High Sierras are relatively young mountains, and the postglacial fell-fields above their tree lines are among the youngest of the landscapes they have to offer. But that's the delight of Robinson's vision. Bringing life to Mars makes it new again. There is no need, in Robinson's world, to give Mars hidden life or ancient secrets or long-lost civilizations to provide some sort of payoff (his books are almost the only major Mars fiction that does not involve any significant discoveries about the planet's past at all). The past is only important to the geologists; the focus is on the planet's future, on its renewal. The flowers of that fell-field in *Green Mars* look new because they are new. They are the latest releases from the genetic engineering workshop of a man named Whitebrook.

As that name suggests, the waters of Mars—the waters that have been at the center of the mythology of Mars for a century and that are central to its scientific understanding today—are at the heart of

Robinson's story of transformation. The details of their release from the depths, and their reshaping of the surface, are related at length, the descriptions lent realism by the learning that so obviously underpins them. The waters are not just crucial to the plot, though; their currents reflect the books' themes and offer a metaphor for their creation.

Like most novelists who choose the epic form, Robinson is deeply concerned with history. The history of a terraformed Mars is an environmentalist history, and water is the most direct and profound of the links between the people and their environment. It is through releasing water that the greatest changes are made; it is water that has the greatest destructive power. At the same time as shaping history, though, the waters also represent it. The waters of Mars transform the planet's surface over the centuries in the same way that the historical forces of politics and economics transform the human world of the settlers. The books' narratives divide and reunite as they run across the decades like anastomosing streams in a flood. The change the waters bring is unpredictable; erosion eats at every accomplishment. But at the same time the waters are the possibility of life; they give the planet's drama an ever-richer range, with ever more surfaces for reflection.

And if what the released waters do to the surface of Mars represents what history does to humanity, it also represents what Robinson's fiction is doing to the science of Carr and all his colleagues. The surface of Mars created by their instruments and mappings is transformed by the imagination that runs over it, left recognizable but altered. Robinson treats the rocks of Mars with the respect due something almost sacred. But the waters are his own. As water transforms Mars, so fiction transforms science.

It's no surprise, then, that water provides the story's basic punctuation. Each volume ends with an attempt at revolution—a change in the flow of history—and with an image of water. In the first, as war encompasses the planet, the water is a vast destructive force, an aquifer breakout tearing its way down the length of Valles Marineris. In the second, the deliberate obliteration of the *ancien régime* corresponds to the slow rise of a flood over the lowland city of Burroughs and the exodus of its inhabitants up the slopes of the

Isidis escarpment to the highlands beyond. And in the third, as a political change more like a crystallization than a revolution takes place, Robinson shows us the level waters that lap against a beach at the edge of Hellas, a new, seemingly stable shore between the sand of science and the sea of imagination. Something at once definite and provisional, as every shoreline made by wind and wave must be.

Shorelines

> The more clearly the immensely speculative nature of geological science is recognized, the easier it becomes to remodel our concepts of any inferred terrestrial conditions and processes in order to make outrages upon them not outrageous.
>
> —William Morris Davis,
> "The Value of Outrageous Geological Hypotheses"

In 1984, shortly after landing a job at JPL, a geologist named Tim Parker found that he had a couple of weeks without too much to do. He decided to spend them in the image facility, looking at *Viking* photographs of Chryse Planitia. As part of a generation of geologists that had entered planetary science after *Mariner* and *Viking*, Parker was quite happy with the idea that vast floods had streamed into Chryse through the outflow channels to the south. What intrigued him was what had happened to the floodwaters after they left the channels. If they pooled as big, shallow, ice-covered lakes before draining into the permeable rocks below, those lakes would have had shores. If so, traces of those shores might still be visible. And if such shorelines were still visible, then Parker's eyes might be peculiarly attuned to their discovery.

Today, North America is considerably better endowed with lake shores than it is with lakes. In the comparatively recent past—during the decline of Earth's most recent Ice Age, between sixteen thousand to ten thousand years ago—the continent was home to many massive lakes that today have more or less completely vanished. Lake Agassiz, stretched out along the edge of the retreating Laurentide ice sheet that sat over eastern Canada, was at its greatest extent four times the area of Lake Superior today. Lake McConnell, to the north of Lake Agassiz, was more than six hundred miles long. On the south edge of the smaller Cordilleran ice sheet to the west was Lake Missoula, roughly the size of today's Lake Ontario. And far from the ice sheets themselves, but fed by the rains that were part of the glacial climate system, were the great lakes of the West: Lake Manlius, part of the bed of which is now Death Valley; Lake Lahontan, in western Nevada; and, largest of all, Lake Bonneville, the shriveled remains of which persist today as Utah's Great Salt Lake. Lake Bonneville has a particular place in American geological history because in the late nineteenth century it became the subject of the great monograph in which G. K. Gilbert deduced the lake's past immensity by measuring its former shorelines—shorelines in the plural, because the lake disappeared in stages, leaving behind the distinctive rugby-shirt striping of a bathtub in a house full of students—and showed how the continental crust had risen up as the vast mass of water was removed.

In the early 1980s the shores of Lake Bonneville again became a site of geological interest. The Reagan administration planned to deploy a new generation of land-based nuclear missiles, the MX or "Peacemaker." Unlike earlier American ICBMs, the Minutemen and Titans (whose descendants sent the *Vikings* to Mars), the MX was being brought forth into a world where the Soviet Union had demonstrated the ability to add new craters to the Earth with great precision and thus destroy missile silos in a first strike. Alternatives to old-fashioned silos were thus needed and one suggestion was that huge missile carriers—some loaded, some empty—could endlessly trundle between various sites in the western deserts on a specially developed network of roads or railways, thus confounding Soviet spy satellites. A great deal of surveying work was carried out in aid of this intercontinental shell game and Tim Parker was one of the surveyors, enjoying what for a geologist was almost heaven—a great deal of time in the field and terrific backup in the form of aerial photography. He spent two years tramping the shores of Lake Bonneville, learning to correlate the subtleties of lake deposits on the ground with the traces seen from the air.

Cloistered in JPL's image facility, Parker started to look for shorelines in Chryse Planitia. Like those who had looked before, he found nothing. If anything, there were hints that the floodwaters had passed through Chryse without stopping and ended up farther north. North of Chryse were the even lower lands of Acidalia Planitia, where strange polygonal fractures in the ground that might reflect the drying up of watery sediments had been seen. But the *Viking* images of Acidalia were not so good, so Parker scuttled across the plains to the east. Here he studied a stretch of tablelands on the edge of Arabia Terra called Cydonia Mensae (the area where the "Face on Mars" had been seen). And there he saw traces of what looked like wave erosion around some elevated islands—islands like those whose shores he had walked and driven over in the Utah desert. Nearby he saw what looked like sand bars, another fossilized shore feature. Soon he was looking at a whole array of what he took to be shoreline features.

Having found possible shores, he set out to discover how big an area they enclosed. But the shorelines did not seem to want to meet up and define lakes. Instead, they obstinately headed north and east along the edge of Arabia's cratered surface into the fretted terrain of Deuteronilus Mensae. There the new features changed their aspect, becoming sharper lines near the feet of the escarpments that mark the edge of the highlands, often facing each other across fretted channels. But they still looked like shorelines; just shorelines cut into a different sort of landscape. More worrying than the change in the features' appearance was the fact that, east of Deuteronilus, they vanished. Parker had come to the blanket of debris surrounding a quite large and very spectacular crater called Lyot that sits just to the north of the northernmost part of Arabia's great curve. On maps it looks like a marble that can't decide whether to roll to the east or the west, or like one of the boulders perched on the rounded peaks of the Marabar Hills in *A Passage to India.* For hundreds of miles, there was no shoreline to be seen. And then, east of Lyot, in the area called Protonilus Mensae, the traces reappeared. At this point Parker started to get really excited. These shorelines were not going to close in on themselves any time soon. This was not the edge of a big lake. It was the edge of an ocean, one that must have filled most of the great northern basin.

In the following months Parker slowly traced his shorelines farther and farther around the highland-lowland dichotomy. The nature of the markings changed from place to place, as it had between Cydonia and Deuteronilus, but Parker knew that shorelines look different in different places. In some places the *Viking* pictures were not clear enough to pick up anything at all. In others, especially around Tharsis, the shorelines became obscured, quite possibly by overlying lava flows, and the thread girdling the ocean was lost. But it could always be picked up again later on. Eventually Parker traced the shorelines down the eastern edge of Tempe Terra, across the southern bight of Acidalia from Deuteronilus, until they finally faded out among the flood features of Chryse. What had started as a couple of weeks indulging curiosity had become the most laborious of circumnavigations, a trip Stan Robinson's settlers would have been proud of. (In fact, Frank Chambers makes a similar trip in *Red Mars*, but in the opposite direction and only halfway around the planet.)

Like earthly oceans, the ocean Parker thought he saw often had more than one shoreline at any given place. This is not that surprising. Changes in the extent of continental ice cover mean that the Earth's seas are surrounded by fossil shores above and below the current ones. During the next Ice Age, today's shores will be anything up to a hundred yards above sea level. In some places on Mars, Parker saw half a dozen shorelines parallel to each other. Most of them, though, could not be traced for long distances. At the planetary scale there were just two: "Contact 1," which kept close to the highland-lowland boundary, and "Contact 2," which was farther out into the lowlands. Parker saw the two contacts—to call them shorelines in print would have been a breach of geological mappers' objectivity—as evidence that the Martian ocean, like Lake Bonneville, had died in stages.

What Parker was seeing was dramatic enough to be slightly scary. At a time when the idea that Mars had significant amounts of water still required championing from people like Carr, to claim that there had been an ocean's worth sitting on the surface would be highly controversial. Parker was by no means a senior figure—he had only recently joined JPL and as yet had no advanced degree. And his evidence was largely the evidence of his own eyes. Other people could look at the same *Viking* frames and see things very differently. Parker, a big but not overly self-confident man, whose loudness is revealed more clearly in his choice of T-shirts than in his blowing of his own trumpet, was not the sort to try to mount a revolution under those circumstances. He proceeded with caution. First he won over the man who had brought him to JPL, Dave Pieri, who had made the first catalogue of Mars's valley networks in the 1970s. He then started work on his superior, Steve Saunders, a *Mariner 9* veteran by this stage more interested in the *Magellan* mission he was planning to Venus than in Mars, and initially quite skeptical.

In 1986, Parker made a presentation on the theory toward the end of a symposium in Washington, D.C.; he overran his time—Parker's details often overflow the limits of his presentation—and thus faced little questioning. Unfortunately he was followed by a man named John Brandenburg, also talking about an ocean but in a much more speculative way, and laboring under the disadvantage that he was also known to be an advocate of the Face on Mars, which the ocean would have made beachfront property (in Face mythology, which draws on the tropes of water shortage and lost civilizations in a reasonably predictable way, the impact that created the crater Lyot is linked to the loss of the planet's atmosphere and ocean). Listening to Brandenburg, Parker recalls, "I was just sinking into my seat." This was the sort of association he least wanted. For the next couple of years he kept quiet and worked on getting his Master's degree.* When his interpretation of the features he saw as shorelines was finally published, in 1989 (by which time his boss, Saunders, was convinced enough to be a co-author on the paper), it was under the wonderfully anodyne title "Transitional Morphology in West Deuteronilus Mensae, Mars: Implications for Modification of the Lowland-Upland Boundary." But the map on the second page showed the shorelines—sorry, "contacts"—stretching all around the great triangle of the northern plains. It was the sort of map people would notice.

One of those who noticed was Victor Baker of the University of

*Parker's caution was echoed by Baerbel Lucchitta in Flagstaff, who also started to think about widespread standing water in the northern lowlands in the mid-1980s, working on the basis of landforms within the lowlands rather than shorelines. Like Parker, most of the time she avoided the word "ocean."

Arizona. When Baker became a professor at the University of Texas in 1971 one of the graduate students there was Peter Schultz, the man who would later make the case for polar wandering to Mars. Schultz was getting access to *Mariner 9*'s pictures, and Baker, who had no background in astrogeology, found the newly discovered surface fascinating. He was particularly interested in the outflow-channel flood features—since quite by chance he was an expert on their closest terrestrial analogue, the "channeled scablands" near the city of Spokane.

The scablands lie on a gently sloping plateau in eastern Washington State, covering an area of about 12,000 square miles between the Snake River to the south and the Columbia River to the north and west. Across this plateau are great dark scars that widen and narrow to some hidden rhythm of the rock, joining and dividing like tangled hair. In places they are interrupted by streamlined islands; elsewhere they are crossed by long cliffs; their beds are grooved and in some places shaped into great long ripples, hard to see except from an aircraft, remarkably gentle in aspect until you remember that they are carved directly into the bedrock.

In the summer of 1922 the geologist J Harlan Bretz asked himself what could have made such a mess of this piece of basalt the size of Belgium. The answer he came up with was water, in very large amounts. By 1923, he was convinced that truly catastrophic floods had swept across the plateau, creating massive analogues to the features often seen in smaller streams—anastomosing channels miles wide, scouring by boulders instead of gravel and so on. Unfortunately, the uniformitarian ethos of geology made such an unheard-of event a very suspect explanation: No one had ever seen such a flood. At a discussion in the Cosmos Club in Washington, D.C., in 1927, Bretz's interpretation was denounced by many of his eminent superiors; few of them encumbered by any practical experience of the area. They all agreed that Bretz's evidence was fascinating; then they all more or less asserted that the Brobdingnagian features must nevertheless have been made by repeated fairly commonplace floods of the Columbia River.

Bretz persisted. He sought ever more evidence but also took a stand of principle, invoking the ideas of William Morris Davis, one of the leading American geologists of the day, who the year before had had a paper published in the journal *Science* championing the role of the "outrageous hypothesis" as a spur with which to prod geology's increasingly settled opinions. Bretz pleaded eloquently to his profession for a fair hearing:

> Ideas without precedent are generally looked on with disfavor and men are shocked if their conceptions of an orderly world are challenged. A hypothesis earnestly defended begets emotional reaction that may cloud the protagonist's view, but if such hypotheses outrage prevailing modes of thought the view of antagonists may also become fogged.
>
> On the other hand, geology is plagued with extravagant ideas that spring from faulty observation and misinterpretation. They are worse than "outrageous hypotheses," for they lead nowhere. The writer's Spokane Flood hypothesis may belong to the latter class, but it cannot be placed there unless errors of observation and direct inference are demonstrated.

Over the next decade the fogged antagonists did begin to look for errors of observation and inference in Bretz's work, even as Bretz himself tightened up the case. Then, in 1940, Joseph Thomas Pardee presented compelling evidence that toward the end of the most recent Ice Age Lake Missoula in eastern Montana had been drained at a truly spectacular rate when a glacial tongue of the retreating Cordilleran ice sheet had suddenly given way in northern Idaho. Bretz had never previously had a satisfactory explanation for the source of the floods and insisted that their existence should be accepted or rejected purely on the basis of the facts in the field—an in-your-face version of "what has happened, can happen" that, while logically coherent, did little to help his case. Now Pardee was giving him the perfect source: thousands of cubic kilometers of water suddenly unleashed through a valley just above the north end of the scablands. In his late sixties, Bretz went back into the field and, with two colleagues, wrote the definitive monograph on the subject, showing that there had in fact been a series of floods, after each of which the ice dam at Pend Oreille grew back, blocking the drainage channel and allowing Lake Missoula to refill itself. In 1979 the Geological Society of America gave Bretz its highest honor, the Penrose Medal; he was ninety-seven.

The citation that came with Bretz's medal was written by Victor Baker. Baker had first come across the scablands when, as a schoolboy, he had made a plaster of Paris model of Washington State. A decade or so later, as a geology graduate student at the University of Colorado, Boulder, he took the Missoula floods as the topic for his dissertation, his first step to becoming Bretz's successor as the acknowledged expert on the topic. When Schultz showed him the *Mariner 9* pictures of outflow channels he immediately recognized some of the similarities and got in touch with Danny Milton, the member of Hal Masursky's USGS team concentrating on the outflow channels. Together, Milton and Baker wrote a definitive paper on the channels as flood features, a paper that beautifully drew out the similarities with the scablands feature by feature. It was what Baker calls an argument from coherence. While the various individual types of feature in the Martian channels were arguably open to other sorts of explanation, the flood hypothesis explained them all at once; it wove them together into a story.

His work on the Missoula floods shaped Baker's future career as surely as water shapes a landscape. It introduced him to a strange form of scenery that he would go on to find in other parts of the Earth as well as on Mars. It taught him how to look at such landscapes and understand them, how to unify far-flung details into a single explanation through a tutored empathy for the force of the waters. In what was still a largely uniformitarian age the floods gave Baker a feeling for—perhaps a taste for—geological catastrophe. And his exposure to Bretz gave him his very own scientific hero to emulate, along with a profound respect for the power of the "outrageous hypothesis." In 1990, he and a set of younger colleagues put forward one of their own, one that pulled all sorts of odd aspects of Mars—including Parker's shorelines—into a coherent but previously unthinkable whole.

While everyone involved in this outrage would claim to have been led by what the data were telling them, in much the same way that Parker found himself whipped off on a trip all around the planet simply by following his eyes, it's hard to talk to Baker for long without realizing that he had come to a stage in his life where an idea's size and scope, not to mention its outrageousness, could be recommendations in and of themselves. Baker is a powerful speaker and a strong debater; he's passionate about the value of geology and a vehement critic of any marginalization of his field by geophysicists and their like. He has devoted a lot of time to thinking about the differences in the way geology and physics see their worlds, the former through experience and imagination, the latter through abstraction and experiment.

"What makes geology strong," Baker argues, "is the reality of its connection to the world, not the logic and structure of its thinking. In physics the connection to the world is tenuous because it only develops after you have produced your model and you're trying to test it. Geologists immerse themselves first. The closest thing I've seen to this is the aboriginal people in Australia: They go out on the land and they have a sense that the land is their dreamtime story, a story in which these rocks and things are their spirit ancestors so everything is sacred but they know it intimately. A good geologist has the same sense of interrelationship and familiarity, a similar closeness of connection." That sense of connection was what allowed Bretz to see the floods at Spokane and Baker to see the floods on Mars: understanding the landscape as a whole, correlating the features, making them cohere. Baker wanted to use the same sense at a planetary level, to develop a hypothesis that would explain a great number of disparate things not by appeal to geophysical modeling but through the sensibility of the geologist.

Jeff Kargel, a planetary geologist now with the USGS in Flagstaff, remembers watching Baker grope toward such a synthesis while Kargel was a grad student in Tucson. "I remember one of the most bizarre talks he ever gave was at a Mars conference in 1988 or 1989, a talk about a hydrological cycle that was mainly subsurface, a very strange talk, a very interesting talk, an unsettling talk. He was seeing that Mars is on one hand somewhat earthlike, but on the other hand it's peculiarly Martian, unique, different from Earth, and this was puzzling Vic, just as it's puzzled many other people before and since." Not that long afterward, Kargel's work helped Baker come up with a solution.

Although his doctoral work was on the outer solar system, Kargel was interested enough in Mars to have committed himself to looking at every single frame sent back to Earth by the *Viking* orbiters, using the university's copy of the NASA photo archive. In early December 1989, he noticed a sinuous braided pattern in the southern part of Argyre, the basin in the southern highlands second in size only to Hellas itself. At first he thought it was a channel; then he realized that rather than being cut into the plain, the winding feature was raised above it.

Kargel's undergraduate work had been in Ohio, a state once covered by the great Laurentide ice sheet; the Ohio River flows close to what was once its southern edge. From field trips there he was quite familiar with eskers, subglacial features formed when water cuts channels in the base of an ice sheet or glacier and deposits sediment on the rock below, leaving something a bit like an inverted plaster cast of an everyday river channel. Pressure exerted by the glacier, not gravity, drives the flow of the water, so eskers can run up slopes as well as down them. What Kargel was seeing in Argyre looked like a system of such eskers. Searching the neighboring frames, he found other features that might be glacial—sharp ridges that could be carved by ice, lumpy curved features that could be moraines. An argument from coherence was starting to form. Very excited, he talked about what he'd found to Virginia Gulick, another grad student, and one who was actually meant to be studying Mars. She was not convinced, but told him to talk to Baker, her supervisor. Baker quickly got excited; he pulled Kargel, Gulick, and a few more kindred spirits into a little impromptu seminar in the university's Space Imaging Center. At the second of their meetings, someone brought along Parker's innocuously titled shoreline paper and forced copies on everyone. Within days the outrageous hypothesis was taking shape.

You can't build a glacier with water seeping up from underground; you need snowfall and so you need water transported through the atmosphere. The problem that Gulick was working on for her doctorate—a set of channels on the northern volcano Alba Patera that seemed to be caused by water erosion—also seemed to call for precipitation. Precipitation seemed to imply evaporation from some sort of sea or ocean. But the Martian ocean invoked by the warm-wet-early-Mars theory couldn't have lasted long enough to have provided precipitation relatively late in Martian history, which was when the glaciers in Argyre and the channels on Alba seemed to have formed. For one thing, all the carbon dioxide would have been turned into carbonates long before. For another, the persistence of water in the atmosphere over such a length of time would have eroded away all the sharp crater walls in the southern highlands. So Baker suggested that the ocean was episodic. It came quickly, lasted long enough to explain what the grad students were seeing, then went away. And it did so repeatedly (hence the many different shorelines seen by Parker). For intimate blending of the earthlike and the alien, this idea has to score pretty highly; oceans are definitely earthlike, but transience is profoundly alien to their nature.

Baker's story was based on the idea that the flow of heat from the Martian subsurface was highly irregular. On the Earth, convection currents in the mantle pump heat to the surface fairly efficiently. A lack of plate tectonics—at the moment, anyway—on Mars suggests that there's no mantle convection and the heat has to move through solid rock by conduction, which is not a very efficient process (if you want to demonstrate this to yourself, get a long thin piece of rock and hold one end over a flame; you'll put down the other end through boredom long before you put it down through pain). Baker suggested that such inefficiency would lead to an accumulation of heat in the mantle and that after a while—hundreds of millions of years, perhaps—the internal temperature would become high enough to force a huge reservoir of magma to the surface in a great volcanic burp.

This volcanic belching would contain a fair amount of gas, most of it carbon dioxide. This would thicken the atmosphere and produce a greenhouse effect. The vast amount of volcanism would heat up a large part of the crust, melting permafrost and opening aquifers, so water would spew out of all the outflow channels. Evaporation from these floods would add water vapor—also a powerful greenhouse gas—to the newly thickened atmosphere; the warmth of the waters' passing would release carbon dioxide frozen into the plains over which they flowed. As the floods filled the great basin of the north, the surface temperature would climb above freezing. Within as little as a few weeks the waters would rise far enough to lap against Parker's shorelines, their waves topped with fizzing foam as yet more carbon dioxide bubbled out into the warming sky.

From the moment the ocean was created, it would be living on borrowed time. Mars would simply not be able to produce a water cycle that could keep the ocean full or a climate that could keep it liquid. Carbon dioxide would freeze directly into the soils around the south polar cap. Carbonated water that fell in the south as rain or snow would seep deep into the porous rocks rather than running straight back to the sea along accommodating rivers, or being pumped back into the sky by transpiring plants, as happens on Earth. Carbonates would form underground and ice in the rocks would freeze. As the atmosphere grew thinner and drier, the temperature would drop until what remained of the ocean—much depleted by evaporation and by seepage into the porous rocks of its own bed—froze solid, forming the ground ice of the northern plains with its peculiar patterns and polygons. But the slow flow of heat out of the Martian interior would continue and, when enough heat built up for another great burst of volcanic activity, hundreds of millions of years later, the carbonated ocean frozen into the rocks would be recycled to the surface.

In this new view, Mars might never have been really warm, but it would have been intermittently wet through much of its history; the limits imposed by the lack of erosion in the south were circumvented by cutting up a few million years of maritime climate and distributing the pieces over billions of years of history. Martian history, rather than being a matter of steady, slow decline, would have been dominated by slow underground cycles, the ocean's brief appearances marking their visible crests. The cycles had begun in the Hesperian, it appeared, and continued until, well, in principle, today. Perhaps. But as the planet's heart cooled, the wavelengths would grow longer; and with each cycle, something would be lost. Some of the carbon dioxide would be lost for good in the form of carbonates in the crust; some of the water would be lost from the upper atmosphere. And so each new avatar of what Baker and his colleagues came to call Oceanus Borealis would be smaller than the last, each new burst of maritime climate shorter and cooler.

In December 1989 its creators thought the episodic ocean did just what an outrageous hypothesis should. It explained a number of apparently quite separate anomalies, such as the valleys of Alba and the eskers of Hellas, the multiple shorelines of Deuteronilus and the details of morphology that suggested that the channels around Chryse, like their analogues in the scablands, had been used more than once. In a flurry of activity the Tucson team produced a set of abstracts for the March 1990 Lunar and Planetary Science Conference in Houston; they ended up presenting four papers back to back. "It upset a lot of people," recalls Kargel. "It intrigued a few." Parker was intrigued but skeptical—among other things, he saw the shorelines of a basin-bound sea in Argyre where Kargel saw moraines. Mike Carr was intrigued too—but perturbed by the way the theory threw around vast amounts of water with unaccountable abandon. The valleys of Alba might represent very easily eroded volcanic ash being eaten away by hydrothermal sources within the volcano itself. The evidence for glaciers was simply not convincing. And you just couldn't get water out of the permafrost and aquifers quickly enough through the application of volcanic heat alone, let alone get the water back into the highlands with rain—enough rain to half empty the ocean—in a few millennia. Carr had a lot of disagreements with the people he began to call the Tucson Mafia; but he kept talking to them.

The Mafia had answers to some of the critiques, but on others they simply refused to be drawn. What they were offering, they said, was a story to be elaborated on and edited, not a model to be falsified. They didn't know quite how the great episodic heatings worked, though they were happy when, a few years later, data from the *Magellan* mission made it possible to argue that on Venus volcanism came in great planet-wide spasms even more vast than the ones they invoked for Mars. (One of the Mafiosi, Bob Strom, played a key role in developing this interpretation of the *Magellan* data.) Nor could they say exactly where the extra carbon dioxide they needed was stored. But as far as they were concerned, their story of the episodic ocean's source needed not to be precise, just to be plausible.

The heart of their case lay not in the details of heat flow, but in their geological feeling for the landscapes; and those landscapes, they argued, could only be explained by brief spates of watery climate relatively late in Martian history. This pattern of argument was familiar to all concerned from the controversy over outrages such as continental drift and the Spokane floods, where field geologists had pointed to things they could explain only by continental drift while others—not just physicists and geophysicists, but also other geologists—argued that the proposed cause was impossible. Over time, more and more disparate features were brought coherently into the picture; the objections were disproved or just dropped and the original geological insight was vindicated.

But for this to happen, new data were needed. Baker and Carr could argue until they were exhausted—there was a big set-piece confrontation at the American Astronomical Association's Division of Planetary Sciences meeting in 1993, chaired by Carl Sagan—but there was no way for either to compel the other to his point of view. Geological history is full of such controversies, situations where people just see things differently. It is a by-product of the fact that geology is ruled by perception and analogy, not experiment. If anything, planetary geology takes this aspect of its parent discipline's nature to extremes—because the time-honored solution of going into the field and looking again, looking together, trying to see more, is just not available. All you can do is wait for the next mission.

The Ocean Below

> Where Alph, the sacred river ran
> Through caverns measureless to man
> Down to a sunless sea.
>
> —Samuel Taylor Coleridge,
> "Kubla Khan"

And when the next mission comes, all you can do is be surprised. If the *Mariner*s and *Viking*s taught their earthbound students one lesson, it was that Mars missions don't settle debates; they start them. *Mars Global Surveyor* has brought this message home to old sweats and novices alike. The sheer amount of data returned by *MGS* since it entered orbit in 1997 dwarfs the haul from *Viking,* which dwarfed the haul from *Mariner 9,* which dwarfed those of its predecessors. And the data differ in quality, as well as quantity: The *MGS* instruments see the planet in different ways from those on earlier missions; they use different wavelengths, different resolutions, different techniques. In short, there is now more to argue about, not less. But if the instruments on *MGS* have not settled the argument, they have certainly stoked it up. Some argue that *MGS* data have bolstered the idea of an ocean on Mars; others that they have weakened the case.

The *MGS* instrument that did most to push matters oceanic to center stage was the altimeter, MOLA. As Jim Garvin—then on the MOLA team, but by 2001 the chief Mars scientist at NASA headquarters—put it, "MOLA is giving the planet its third dimension. And if you want to know what ends up where, that third dimension is all important." Water flows into holes and once it's done flowing, it defines what geophysicists call an equipotential surface with respect to the planet's gravity field, and what everyone else calls a sea level. Discovering holes and ascertaining whether things are level are both things at which MOLA was designed to excel.

Jim Head, a professor at Brown University, had a pretty busy time of it in the 1990s, working on reams of radar data from the *Magellan* mission to Venus and on data from the *Galileo* probe's encounters with the moons of Jupiter, and in the process getting into the thick of various quite heated arguments. He hadn't had much time to think about oceans on Mars; inasmuch as he had thought about them he found the idea far-fetched. But Head, well aware of the habit space missions have of provoking ever more new ideas, is keen to test hypotheses whenever he sees a way to do it; it is, after all, the only way to cut down their multiplicity. And as one of the members of the MOLA team, he had great new data with which to test Parker's ocean hypothesis. If Parker's "contacts" were shorelines defined by ancient sea levels, then they should be level, and MOLA measurements should demonstrate the fact.

In the case of Contact 1, the higher of the two planet-girdling features in Parker's first paper, the MOLA data were all over the place—the contact wandered up and down enthusiastically, suggesting an ocean unrealistically well suited to water skiing. But Contact 2, farther to the north and lower down, was a different story. All around the planet its elevation changed by less than a thousand meters (3,000 feet). What's more, its departures from the horizontal made sense. The contact went up where the land might have been lifted since the shoreline was etched by the sea—on the flanks of Tharsis, for example. It went down where the land might have been lowered. Contact 2 looked like a pretty convincing shoreline.

Intrigued, Head and his colleagues went on to further tests. The MOLA data could be used to measure the roughness of the surface being scanned and it turned out that this, too, seemed to back up Parker's story. The areas south of Contact 1 were always rougher than the areas north of Contact 1 and south of Contact 2. The areas north of Contact 2 were smoother still. This would fit with the idea that there had been an ocean—sedimentation makes sea floors very smooth. Another test was to use the altimetry to get a sense of how great the volume of an ocean that had lapped up against Contact 2 would have been. The answer turned out to be 14 million cubic kilometers. Undeniably a lot of water, but only 1 percent of the amount in the Earth's oceans; enough to cover the surface of Mars to a depth of about a hundred meters, if spread evenly, but well within the range that Carr had made acceptable. Some regional details seemed to add to the case. For example, Head saw that the erosional features that defined the outflow channels around Chryse all ended at more or less the same elevation, as might be expected if that elevation marked the point at which they flowed into a sea. And that elevation was pretty close to Contact 2.

As well as providing the data with which to test aspects of the ocean hypothesis, MOLA also provided a substantial, if not exactly scientific, iconic boost for the idea. When the MOLA data were used to make maps, colors had to be chosen to represent the different elevations. The color scale that was used went from blue for the lowest of the basin floors through green, yellow, orange, red, and brown to a bleached-out white for the mountain heights. And so MOLA's maps showed a smooth swathe of blue across the north of the planet, while coloring the rougher southern highlands with the sorts of yellows and oranges that might be taken as naturalistic representations of the Martian surface. All that was being shown was elevation—but if anyone looking at those maps did not, now and then, fall into the fallacy of seeing a great blue ocean lapping at the edge of a cratered continent I'd be amazed. I certainly did. I still do.

The maps may or may not have had a subliminal effect. Head's results, published first in late 1998, definitely bolstered the idea of an ocean. And then Parker himself moved the debate forward. At the Lunar and Planetary Science conference in Houston the following March, he presented not just more evidence for shorelines at some point in the past, but a far-reaching model of a long-lived ocean—an ocean that didn't disappear at the end of some early warm, wet period, but which evolved through most of Martian history and perhaps up until the present day.

Parker's collaborator in this new work was Steve Clifford, a planetary scientist based at the Lunar and Planetary Science Institute that hosts the annual Houston meetings. In the early 1990s Clifford published an intriguing model of how water might cycle through the Martian crust, a model designed to explain the evidence that there was some ground ice in the Martian tropics. This evidence was troubling because ice on or near the surface in the tropics would tend to sublime in sunlight. Over time, this process should have removed all the ice from the equatorial soil; but there still seems to be some of it there.

Clifford's argument was that the ice had got there from below. He suggested that deep down in the megaregolith, where the planet's heat would keep things warm enough for liquid water, Mars might have a connected planet-wide aquifer. Above the water table in this aquifer there would be traces of water vapor. That vapor would percolate upward until it reached rocks close enough to the surface to be below freezing, at which point it would freeze to them. The water in the aquifer would be replenished from the polar caps. Since ice does not conduct heat very well, Mars's polar caps must act as insulators, damming the upward flow of geothermal heat from the planet's interior. When they reach a certain thickness the ice at the bottom will start to melt. Melt water would then trickle down into the megaregolith aquifer.

Clifford's model was intriguing and seemed to explain how frozen soil could persist in the Martian tropics by being replenished from below. But it was also rather hard to get to grips with. Clifford's theory was based on hydrological models that used mathematics not necessarily familiar to geologists. And though he was able to evoke earthly analogues for many of the processes involved, they were hardly commonplace. The base of the Earth's Antarctic ice cap melts because of the geothermal heat flux below it, but at the time Clifford's paper came out that was a fairly specialist interest.* Water under Australia and in continental basins elsewhere flows for hundreds, even thousands of miles through cracks and pores in otherwise solid rock, but even in hydrology such large-scale flow was not much discussed until relatively recently. The geological engineers studying the safety of the proposed nuclear waste repository at Yucca Mountain were aware of the way moist vapors could rise

*It's far more widely known about today, thanks to the discovery of a vast pool of melt water, Lake Vostok, two and a half miles below the surface of the ice.

from aquifers many hundreds of feet below, but it was hardly a mainstream concern. A very unearthly water cycle invoked by means of analogies to quite obscure earthly processes was hard for any but diehards to come to terms with. Carr, definitely a water-on-Mars diehard, not to mention an admirer of some of Clifford's work, studied the idea closely; but even he ended up not quite sure what to make of it.

The paper Parker and Clifford presented in 1999 combined Clifford's ideas about aquifers with Parker's ocean. Clifford had imagined melt water from the poles filling the nooks and crannies of the megaregolith up to the same level all around the planet: The water table would be a sort of underground sea level and so it would be flat. The surface of Mars, though, is far from flat. It has a distinct slope, with the south pole about six miles higher than the northern plains. So if the water were to find its level on a planetary scale, that level could easily be deep beneath the southern highlands—and a few miles above the northern plains. The ocean's shore would mark the contour at which the ground dipped below the water level in the aquifer. Northern ocean and southern aquifer would be part and parcel of the same thing.

An ocean would not necessarily mean a warm climate, with clouds and rain and such. Parker and Clifford imagined it to have been covered with ice from close to its inception. But it would still have been able to drive a strange, subdued sort of global water cycle. The top layers of the floating ice would have slowly sublimed and most of the water vapor lost through this process would eventually have frozen out at the south pole. As the southern cap grew, its base would melt, and water would trickle down and north through the permeable crust until it reached the water table. So water that moved from the north to the south through the atmosphere would flow back from the south to the north along uncharted subterranean slopes, effectively rejoining the ocean.

But this recycling system had a leak. In the rocks above the subterranean water flowing north there would be water vapor that would diffuse upward until freezing into cold rock near the surface. And as the planet's internal heat seeped away and the cold of the surface penetrated ever deeper, the bottom edge of that freezing layer would creep lower and lower. The thickness of frozen ground would increase, increasing amounts of ice would be layered on to it from below, so more and more water would be lost to the atmosphere/aquifer/ocean system. As water was locked away in intractable ice, the sea-level/water-table would drop. In its earliest days the ocean might have reached the equator; in 1997 Parker and Ken Edgett had identified what they thought was another shoreline—the highest and oldest to date—in the Terra Meridiani region of western Arabia. But as time went by it would have retreated, its successive shorelines tightening round the northern lowlands like a noose.

To begin with, the recession would have been orderly. But at some stage the sea level would have dropped low enough for a large part of the ice covering the ocean to have frozen to the rock beneath it. Water would still have sublimed off the top of the northern ice sheet and frozen to the southern cap; ice at the base of the southern cap would still have melted and flowed back toward the north. But water would no longer have been able to get into the ocean, because the ocean's surface was locked at a set level by the grounded ice. So the water table and the sea level went their separate ways. The water within the rocks of the southern hemisphere started to back up above the level of the ocean's icy surface. Yet it didn't spill out over the top of the ice, because it was held back by the impermeable layer of frozen ground below the surface. By this point, according to Clifford and Parker, the frozen ground might have been two miles thick or more, and two miles of frozen ground can withstand a lot of pressure. So as water kept flowing down from the south pole the water table would have backed up higher and higher. A lot of pressure would have built up.

Thus the stage was set for the creation of outflow channels by mechanisms very like those that Carr had detailed in the early 1980s. When the pressure got too much, or an impact broke through the frozen ground, torrents of water unable to fit underneath the ground-bound ice sheet spilled out over it instead. Sometimes the sea ice itself would rupture and a vast fountain of high-pressure water would rise into the sky. These sea bursts in the northern ice plains, driven by the pressure exerted by a water table maybe four miles or more above them, might have been far more spectacular than the outflow floods around Chryse. Through them the ocean could have been reborn time and time again—but always to freeze once more almost immediately, always to recede again as sublimation took the water vapor from its surface and the cooling planet stored more and more ice in the cold rocks of its upper crust. In the end, all would be ice, covered by dust and, in places, lava. Whether that end had been reached, though, was hard to say. As Parker pointed out at the meeting in 2000, there is an outflow channel—Athabasca Vallis—in the terrain that Alfred McEwen and his colleagues think they have shown to be very recent. Regional aquifers capable of producing floods might thus persist to this day.

The model Clifford and Parker unveiled at the Lunar and Planetary Science Conference in 1999 thus explained why the outflow channels did not appear until much later than most of the valley networks and other signs of water on the surface: Before the sheet of sea ice over the ocean was thick enough to stick fast to the rocks below and around it, the pressure needed to drive the outbursts could not build up. It also explained why they had tapered off: As the layer of frozen ground got thicker and the level of water in the aquifer lower there was ever less of a chance for the water pressure to break through. It explained shorelines; it explained ground ice. It was an ambitious and far-reaching account, in that it provided a framework for thinking about the evolving roles of water and ice throughout Martian history, rather than just in the putative early warm-wet period. But at the same time it was not too terribly radical in its implications. It sat fairly happily with the basic elements of the Mars Carr, Baerbel Lucchitta, and others described in the 1970s and 1980s, a Mars with big thick layers of ice and sediment in the northern plains, a Mars on which surface water has played only a small role. With its periodic resurgences it might even capture some of the attractive points of the Tucson Mafia's model.

By the time the community came back to Houston for the following year's conference, the frozen ocean was a hot topic. Jim Head convened a "microsymposium" to debate the evidence—and even Mike Carr had to admit he was impressed. There were sessions in the meeting proper on the implications of oceans; there were enough abstracts on the subject for a whole section of the conference's poster session to be entitled "Blue Mars," in homage to Stan Robinson.* Vic Baker gave the meeting a characteristically combative restatement of the episodic ocean theory, which he had started to call the MEGAOUTFLO (Mars Episodic Glacial Atmospheric Oceanic Upwelling by Thermotectonic FLood Outbursts) hypothesis. But the idea of a permanently frozen ocean was raising interest among some of the Tucson Mafia.

More fuel for the debate was coming from the *MGS* Thermal Emission Spectrometer (TES). If people had been polled in advance of the *MGS* mission as to which instrument might have the most bearing on matters oceanic, it's a fair bet that TES would have been the instrument nominated. The canonical warm-wet-early-Mars story requires large beds of carbonate formed in standing water on the surface to have gobbled up the atmospheric carbon dioxide. Carbonate beds would mark out the floors of ancient seas and oceans like vast flat tombstones. But when TES set to work, the carbonates failed to turn up. As far as TES could make out, the whole planet was covered with minerals typical of basalts and other products of volcanism. The only large exception was that in part of Terra Meridiani the spectrometer picked up what appeared to be evidence of big crystals of hematite, an iron-bearing mineral. On Earth, hematite in this particular form would only be expected in dried-up lakes or hydrothermal systems. On Mars it was sitting in what Parker would say had been the coastal shallows of the northern ocean's greatest, earliest extent—a connection the TES team was happy to make clear in its presentations on the subject.

*It's worth mentioning that although Mars scientists on the whole seem rather pleased by the fact that Robinson cares enough about their work to have researched it thoroughly and used it extensively, surprisingly few seem actually to have read his Mars books in their entirety. Some have told me that they already know about Mars; others that Robinson makes exploring the planet too easy for his characters. Some may just not like the books' style, or their politics, or their admittedly daunting length. Though there are doubtless others, the only practicing Mars geologist I know to have read all three volumes is Jeff Moore, at NASA Ames, who read them more for their utopian, anticapitalist take on the practice and politics of science than for the science itself: a deacon in the Church of the SubGenius with an academic background in European history and literature, Moore is rather more countercultural than the average planetary scientist.

None of this was conclusive, or even truly compelling. It certainly wasn't coherent. There are obvious differences between the ocean Baker talks about and the ocean Parker talks about. The hematite could have been made long ago by a small body of water in a warm wet world that had since vanished, or—as is now thought likely—by a hydrothermal system. It could have been made without recourse to water at all. Ken Edgett on the MOC team was deeply dubious about a great deal of what was going on. Despite the fact that he had written the original paper on the putative shoreline in Terra Meridiani with his friend Parker just a few years before, and had worked with the TES team at Arizona State University, Edgett was unconvinced by both their contributions. He had come to think there was no shoreline in Terra Meridiani; indeed, by 2000 he no longer believed in shorelines anywhere on the planet. He and Mike Malin had used MOC to look at sections of Parker's contacts and they saw no features that they can interpret as shores. Unsurprisingly, Parker's interpretation of the new images differs.

But throughout 2000 watery interpretations kept coming along with every new bit of data. A new meteorite study suggested that the atmospheric deuterium/hydrogen ratio—frequently taken as evidence that Mars had lost water to space—had in fact been high from the beginning, and so the planet's reserves of water might either be larger than had previously been assumed, or more regularly cycled through the atmosphere. Another meteorite study showed vestiges of salty water in the far-flung rock, suggesting it had been bathed in oceanic brine at some point. And Edgett himself found data that dragged him, "kicking and screaming," to the announcement that whether or not there were any oceans in the past, there was definite evidence that liquid water was occasionally flowing across the Martian surface right now.

"Common Sense and Uncommon Subtlety"

"Ice is cool. It's water, but it's not."

—Oz in *Buffy the Vampire Slayer*
("Helpless" written by
David Fury)

Nathalie Cabrol is an intense, cheery and above all animated woman bewitched by the possibility of lakes on Mars. She started working on ways of identifying craters that might once have held water as part of her degree work at the University of Paris; she and her partner, Edmond Grin, now continue their work at NASA's Ames Research Center. For years they have studied the complex history of a crater called Gusev, a basin in the edge of the highlands southwest of Elysium (it's just big enough to pick out on global maps at 175° E, 15° S). A large valley, Ma'adim Vallis, drains into it; there are signs of a river delta, of several shorelines, of a flood breaking over the northwest ramparts, maybe of ground ice. Working through all this evidence, Cabrol and Grin have charted Gusev's history as an ice-covered lake through a substantial fraction of Martian history.

In the late 1990s, Nathalie Cabrol and Edmond Grin set about creating a gazetteer of all the crater palaeolakes on Mars—every crater that had clear evidence of something, sometime, having drained into it. They found about 170. Most of them were Hesperian and scattered around the planet's equator. This made sense: Since the Hesperian was when the outflow channels were active, some of that water might have pooled in lakes directly after its release. And some water would have evaporated and ended up as snow—or even, if the outflow floods were giving off enough carbon dioxide, rain—to fill lakes far away.

Some of the lakes, though, were more recent, with sediments that seemed to have been laid down in the Amazonian. These lakes were not so numerous, nor so widespread. They were grouped in a set of clusters: They were at the south end of Amazonis, around the edges of Chryse, in the Argyre and Hellas basins, and in a few patches in the northern parts of the highlands. One day in late 1999, shortly after mapping out these sites, Cabrol went over to one of the other space-science buildings (the offices at Ames are scattered around a site dominated by vast wind-tunnels and yet vaster airship hangars) to see her colleague Robert Haberle. Though Haberle's office door bears a sign proclaiming him "Commander of the Solar System," his primary interest is the Martian climate; he builds computer models of it and longs to improve those models with data from a system of small, long-lived weather stations, if he could only get someone to pay for them. Cabrol was doing some work on the possibility of future small missions to Mars and thought they might be able to help each other. When she got to his office, though, technology flew from her mind. There was a map on his desk that showed the sites where, according to Haberle's model, the atmospheric temperature and pressure might be high enough today to allow a thin dew of liquid water for a few midsummer days every Martian year. And Haberle's map looked remarkably like the map Cabrol had just left on her own desk.

Cabrol rushed back to get her map so that she and Haberle could compare the data in more detail. They both got excited, Cabrol slightly more so, perhaps due to temperament, perhaps because Haberle works with climate models a lot and thus has a feeling for the degree of credence their suggestions merit. If the model was on the button, though, it seemed to be saying that the places where Cabrol saw the best evidence for water in the recent past were the places where there was the greatest likelihood of water today. Other results from Haberle's model suggested that if there was a lot of ice in the northern plains, a change in the planet's obliquity that subjected those plains to long, uninterrupted periods of summer sun would put enough water vapor into the atmosphere to feed snowfall elsewhere on the planet. To Cabrol it started to look as though, regardless of whatever might be going on in any putative aquifers, the planet still had traces of a hydrological cycle in its atmosphere that might account for the occasional ice-covered lake. And research in the Antarctic, some of it by colleagues at Ames, had showed that water could persist in ice-covered lakes for a very long time, provided their surfaces were heated above 32°F for a few days every year. Under the dust and ice some of the craters might conceivably have thin layers of liquid water in them today. Mars might be hydrologically dozy, but it wasn't hydrologically dead. Water on Mars was not reserved for the warm, wet past; it had a cold, clammy present.

Haberle and Cabrol gave a tag-team set of papers at the following spring's Lunar and Planetary Science Conference, the meeting that also saw the microsymposium on oceans. The audience was thrilled. Virginia Gulick and Jeffrey Kargel, sometime Tucson Mafiosi, found Haberle's work particularly fascinating. During periods of high obliquity Haberle's model put snow on Tharsis, not too far from the relatively fresh new channels Gulick and Vic Baker has studied on the flanks of Alba Patera. Snow might also be expected in Argyre and Hellas, the places where Kargel thought he'd spotted glacial landforms. Since that meeting Kargel has become even more convinced that the features he found in Argyre at the end of the 1980s really are glacial landforms, but he's no longer sure you need brief bursts of a semimaritime climate to create them. Instead, he's pursuing the possibility that the glaciers might be built up over ages by very low levels of precipitation like those suggested in Haberle's models; the main reason he thinks this is possible is that he's sure at least one of the high-resolution MOC images of the area shows a glacier moving across the surface right now.

That's not the only thing the MOC images show. One of the very earliest of them revealed what appeared to be something seeping down the sides of some of the craters in the south. Malin showed the image at a press conference and many people found it intriguing. Nathalie Cabrol was particularly excited, especially once she found other evidence for an almost contemporary hydrological cycle. But it was only after the orbit was finalized and the resolution considerably improved that Malin and his colleagues appreciated quite how important these seeps and gullies might be. While almost every steep slope on Mars shows signs of something having slipped down it, the new gullies, which had been seen in about a hundred different places by early 2000, were quite distinctive. At the top of the gullies were "alcoves" where the overlying rock and regolith had been undermined; below were channels where water seemed to have assisted the flow of debris down the slope. Beneath these channels were aprons of debris from previous flows. The gullies almost always came in groups, all their sources coming from the same layer of rock in the wall of the crater—or, in some cases, valley. And they were very fresh indeed. None of them was blemished by a single impact crater; some of the aprons rolled out over what appeared to be active dune fields. They looked really, really recent.

After *Viking*, Mike Malin had spent a lot of time going on field trips to places he thought might look Martian; he'd seen gullies like this on the edge of the Colorado Plateau, in places where water flows along a permeable stratum of rock to an otherwise arid cliff. He became convinced the Martian gullies were the result of similar seepage in the walls of the various craters, valleys and pits where they were seen, and converted the skeptical Edgett. They sent off a paper for publication in the journal *Science* and, when it was accepted, alerted NASA to the fact; the result was a press conference at NASA headquarters, the first on a scientific result concerning Mars since the one held in 1996 to announce the possibility of evidence for life in the Martian meteorite ALH 84001. Such a fuss was hardly surprising. Water on Mars had been chosen as the overarching theme for the "Mars Surveyor" program that *MGS* was meant to have begun. Since the rest of the Mars Surveyor program had by this stage either crashed or been canceled, a striking success was a nice thing to brag of.

Mike Carr, a member of the MOC team, was brought along to the Malin and Edgett press conference on the quite reasonable basis that he literally wrote the book on water on Mars. He described being simultaneously drawn by the strong geological evidence in the pictures of the gullies and appalled by their apparent physical impossibility. The seeps they were seeing were about a hundred yards or so down the walls in which they appeared. Rocks at such depths would be almost as cold as the surface—and far too cold for liquid water. These seeps were not, after all, in the tropics. They were all at latitudes of 30° or more, mostly in the south, and they faced toward the pole, thus receiving a minimum of direct sunlight. The sites of the seeps were cold even by Martian standards, a long way from the balmy climes where Haberle's climate model allowed surface water. Some were quite close to the south polar cap, in near permanent shadow.

If water was seeping to the surface in these gullies, even extraordinarily salty water with a freezing point far lower than normal, then either the interior of Mars was much warmer than expected, which would mean its surface was a bizarrely effective insulator, or the climate must recently have been a lot warmer than it is today. If either of these things were the case, you'd expect a lot of other watery features. But they weren't there.

So maybe, Carr suggested, this story wasn't about water. Under some circumstances, water ices can trap molecules of gas within them to form gas hydrates, also known as clathrates, which have physical properties quite distinct from those of pure ices. Clathrates can release their stored gas quickly in response to temperature and pressure changes; and when a clathrate breaks down in this way, the gas released will occupy a far greater volume than the clathrate itself did. Sitting on the panel at Malin's and Edgett's press conference, Carr told the reporters that a layer of clathrate giving out sudden bursts of carbon dioxide might be an alternative explanation for the gullies. In a commentary that accompanied the Malin and Edgett paper in *Science*, Ken Tanaka of the Survey said much the same thing.

An increased planetary role for carbon dioxide has always been the great alternative to Martian earthliness. In the nineteenth century it was suggested that the Martian ice caps were not made of water ice but of carbon dioxide, which freezes at a much lower temperature (–110°F) and would strongly suggest the planet was too cold to be habitable. Alfred Wallace, co-enunciator with Darwin of natural selection, argued strongly for carbon-dioxide caps in his attempts to refute Lowell's picture of an inhabited Mars (Wallace liked a cosmos in which humanity might be unique). Lowell would have none of it, seeing the idea as an ingenious piece of special pleading designed merely to make the Earth seem exceptional:

"Had half the ingenuity been expended in testing the theory as in broaching it," he chided, Wallace and his like would have realized that because carbon dioxide only becomes liquid at moderately high pressures, carbon-dioxide caps would sublime directly into the atmosphere when the weather warmed up and thus could not leave the ring of moistened Earth that telescopes seemed to show around the caps during the Martian spring. Faced with "the rival candidates of common sense and uncommon subtlety," Lowell favored the Copernican common sense of frozen water, a position that allowed him the earthlike Mars he wanted.

When Robert Leighton and Bruce Murray predicted in 1966 that the caps were in fact largely carbon dioxide, and *Mariner 7* corroborated the idea in 1969, uncommon subtlety made a comeback. But when valleys and channels were seen all over the planet and the north pole was found to contain significant amounts of water-ice, a commonsense, more earthlike Mars reasserted itself, at least in the planet's past. Danny Milton, the *Mariner 9* television team member who wrote the first papers on the outflow channels, briefly straddled the two camps with the suggestion that carbon-dioxide clathrates might provide the channel-carving floods with their unearthly power. The breakdown of underground clathrates heated by volcanism would not only produce liquid water, Milton speculated. It would also produce a lot of gaseous carbon dioxide to help blow that water out of the ground—"a peculiarly Martian explanation of a peculiarly Martian feature." But this explanation didn't stick. After Mike Carr had explained outflow channel floods through the rupturing of aquifers, there was no longer any need to invoke melting of permafrost, either with clathrates or without. Indeed, despite the fact that there was a lot of carbon dioxide and ice on Mars, the idea that carbon dioxide might play a role in shaping the planet's surface, either directly or by modifying the behavior of frozen water, fell by the wayside. The massive compendium of papers on Mars published by the University of Arizona in 1992 doesn't have the word clathrate in its index; as far as I can see, no paper on Mars published in the 1980s mentioned them at all.

By the late 1990s things had changed. Even before Malin's gullies were revealed in the summer of 2000, clathrates were getting a fair bit of attention: At the Houston meeting that spring, Jeff Kargel and Steve Clifford had both presented papers on the roles clathrates might play in the subsurface. More dramatically, Nicholas Hoffman, something of a voice in the wilderness (especially since he is based at La Trobe University in Bundoora, Australia), had put forward a mirror image of conventional thinking on the roles of carbon dioxide and water on Mars. He suggested that throughout Martian history the planet's reserves of water have remained almost entirely frozen, some of it as pure ice, some as clathrates, and that the pore spaces in the megaregolith where Clifford and Parker would put the planet's water in fact contain liquid carbon dioxide. Floods of boiling carbon dioxide, not water, are to be held responsible for the outflow channels. This model—which Hoffman calls "White Mars"*—has few adherents as yet. Some think Hoffman doesn't know how goddamn wrong he can be (indeed, some people think that about all the notions in this chapter). Others find his ideas thought provoking, if extreme. Talking to me about Malin's and Edgett's gullies, Robert Haberle pointed out that the strata from which they emerge are just far enough below the surface to offer the sorts of pressures—five Earth atmospheres—you need to liquefy carbon dioxide. Ken Tanaka and Jeff Kargel in Flagstaff have started working with Hoffman.

Why have carbon dioxide and clathrates made a return? There are three possible answers, all of which probably played a role. The first is that Martian geology has taken an icy turn and thinking about ice

*There is also a novel of the same name by Roger Penrose and Brian Aldiss, intended as an antiterraforming alternative to Stan Robinson's books. As to further colors: Larry Niven has written a book called *Rainbow Mars*; Geoffrey Landis, who works on solar-cell technologies for NASA and was a member of the *Pathfinder* team, has written a novelette originally called *Brown Mars,* though it was published under the title *Ecopoiesis;* there is a volume in the interminable "New Adventures" series of *Doctor Who* spin-offs called *Beige Planet Mars; Black Mars* is a book about the role of African mercenaries in Attic Greece by Desmond Bernal. (All right, I made that last one up.) Robinson spoofs his own style and concerns in a sweet little parody called "Purple Mars," to be found at the end of his book of short stories, reflections, and poems *The Martians,* which also includes the original novella *Green Mars.* At the time of this writing, the nicely alliterative "Magenta Mars" and "Mango Mars" are still up for grabs.

seriously leads one to think about clathrates, too. One part of the icy turn can be traced to the fact that in the late 1970s and the 1980s JPL's *Voyager* probes sent back images of truly icy places such as Jupiter's moons Ganymede and Europa, and Neptune's moon Triton, places where ice has to be understood not as a form of water but as a solid mineral making up a large part of the satellite's crust. Another part of the story is fieldwork. In the 1970s very few Martian geologists had spent time in the Earth's Arctic or Antarctic; now many have, drawn by a desire for the analogues to Mars that might be found there. Iceland, with glaciers and volcanoes in close interaction, has become a Mecca for Martian geology; Reykjavík saw two conferences dedicated in whole or in part to Mars in the summer of 2000. Some Mars experts have now spent a lot of time in the dry valleys of Antarctica, where among other things they were able to understand how ice-covered lakes might keep their depths liquid for millions of years on the basis of a few days above the freezing point every summer. Over the past few years Mars geologists have started to visit the polar deserts of Devon Island, in Canada. There they've found streams formed under ice cover that follow patterns very like those of Martian valley networks; they've seen how snow stuck on slopes can create gullylike features when it melts without the help of any subsurface water at all. Mars geologists are developing a taste for such landscapes. In this context, thinking about clathrates on Mars is just an extension of thinking carefully about ice and how it behaves.

The second explanation is that clathrates might solve a real problem. Venus has an atmosphere ninety times as massive as the Earth's made of almost nothing but carbon dioxide; the Earth has two-thirds that amount of the stuff locked away in carbonates. The lack of a thick carbon dioxide atmosphere on Mars suggested that it has a similar stash, laid down in the warm-wet-early period. However TES, the *MGS* spectrometer, sees no sign of those carbonates. The normal response to their absence is to suggest that they are buried, perhaps under fairly thin layers of dust; give Mars a good going over with a broom and the carbonates will turn up. But if Mars's water was always frozen, or sealed away from the atmosphere under sheets of ice, then even if it was available in copious amounts it might not have produced much in the way of carbonates. If the carbon dioxide isn't in the atmosphere and isn't in carbonates, a new place to store it is needed. Clathrates offer one.

Clathrates are not just a handy place to deposit surplus carbon dioxide; they're good for making quick withdrawals when the stuff is needed in a hurry. It was the Tucson Mafia that reintroduced clathrates to the Martian debate and they did so because their episodic ocean required large amounts of carbon dioxide with which to form a greenhouse more or less on demand. Vic Baker—who had worked with Danny Milton on the outflow channels back in the 1970s and thus knew his clathrate theories—calculated that clathrates in the permafrost might contain enough carbon dioxide to make the Martian atmosphere about six hundred times thicker than it is today. At the time, this sudden invocation of an idea widely ignored for a decade could be seen as a little desperate. Ten years on, though, with TES failing to find any carbonates on the planet's surface, the idea of underground clathrates storing large reserves of carbon dioxide has rather more in its favor, though many remain extremely skeptical.

The third reason for an interest in clathrates might be as simple as fashion. In the early 1970s, naturally occurring methane clathrates had just been discovered on the Earth's sea floors. Those early studies were doubtless on Milton's mind when he gave carbon dioxide clathrates a role on Mars. After the thrill of discovery, though, earthly clathrate research was fairly low-key stuff and easy to ignore. Then, in the 1990s, it took off. The decomposition of methane clathrates beneath continental shelves came to be blamed for massive collapses of the sea floor. The methane given off by clathrate decomposition began to be talked about as a putative cause for catastrophic climate changes in the distant past—or possibly in the not-so-distant future. Clathrates even developed potential economic importance: They may be by far the largest single source of hydrocarbons on the Earth. By the mid-1990s Earth scientists were pretty excited about clathrates; it was only natural that scientists studying Mars should take another look to see if they might make a difference there too.

The idea that theories about Mars reflect trends on the Earth rather than new data from spacecraft might seem depressing; space travel loses some of its exploratory excitement if its purpose is merely to hold up distant mirrors to trends in terrestrial thought. But something bigger is afoot. In the decades since the space age began, and in a process that must surely owe something, though not everything, to the insights gained from planetary exploration, uniformitarianism has increasingly lost its grip on the science of the Earth. The processes that shape the Earth have more and more come to be seen as evolving over time, sometimes gradually, sometimes catastrophically. The Earth's history has become a site for speculation and grand theory as well as the careful back projection of analogies from the present. When *Mariner 4* sent back its first images of Mars, the plate tectonic revolution was barely begun; the link between the Earth's Ice Ages and its orbital oscillations was unproven; hardly anyone took seriously the idea that asteroid and comet impacts played a catastrophic role on the Earth, causing mass extinctions like that of the dinosaurs. Vast belches of methane from sea-bottom clathrates changing the climate, "snowball earth" glaciations reaching all the way to the equator—such alien events were not on the agenda. Now such things are at the core of the Earth sciences. The realm of the "earthlike" has been vastly expanded.

Mars may well have had a warmer and wetter past. But it might not have been as warm and wet as analogies to the Earth made it seem in the 1970s. Perhaps the ocean was always covered with ice, as the snowball theorists think the Earth's once were; perhaps valleys were laid down under thin sheets of snow, as they are in the Arctic; perhaps the channels were carved by great streams of ice, as Baerbel Lucchitta has been arguing for twenty years—vast ice streams seem to have carved similar patterns into the sea floors around Antarctica. An enriched view of the Earth has expanded the possibilities for Mars.

Such speculations carry with them a new view of Mars in which the commonsensical and the uncommonly subtle are bound together to make something new, like water and carbon dioxide in a frozen clathrate. Martian history need no longer be divided into two parts, early and earthlike versus late and alien. It is showing signs of becoming a single strange story. It has a dying fall, with hydrological activity becoming weaker and weaker just as the planet's volcanism has. But as with the volcanism, it may well not be over yet. There may still be water moving around today. Theories about the depths of the megaregolith, or hidden ice-covered lakes, might never convince anyone of that on their own. But something made those gullies and, though some uncommon subtleties may have played a role, common sense really does insist that water had something to do with it.

Water on Mars matters in ways that go far beyond the geological. Water defines worlds in myth and metaphor because water separates the living from the dead. And this seems to be the case in science as well; as far as we know, water is a prerequisite for life. Another part of the explosion of knowledge in the Earth sciences over the past few decades is the realization that life can persist under conditions that humans would consider hugely inclement, as long as there is liquid water available. To look for life on Mars is still a long shot, but if there really is some wetness still within it's possible that it might pay off. If only we could get there reliably. If only we could find the right place.

Part 4 – Places

To any observer, it would have been an interesting demonstration of the slowness of some mental processes. For both Gibson and Jimmy had walked a good six paces before they remembered that footpaths do not, usually, make themselves.

—Arthur C. Clarke, *The Sands of Mars*

Buffalo

It is not down on any map: true places never are.

—Herman Melville, *Moby-Dick*

It is 22 June 1999. On the Earth, the Northern Hemisphere's summer solstice has just passed; on Mars, the southern hemisphere's spring is shortly to begin; between the two planets and independent of all seasons, *Mars Climate Orbiter* and *Mars Polar Lander* are swinging outward to their separate dooms. And in the Audubon Room of the University Inn, Buffalo, it is happy hour. The planetary geologists gathered around the tables in one corner are aware of all these facts, but it is the last that occupies them most. They have not, for the most part, started off very happy, but they are making full use of their entitlement to two drinks for the price of one, or eight for the price of four. There is shooting of the breeze and the occasional whiff of flirtation; a few chairs are fallen off, a few glasses broken. And the frustrations of the meeting that has just wound up are thoroughly chewed over.

The meeting in question was the second site-selection workshop for the 2001 Mars Surveyor Lander, which was held at the Buffalo campus of the State University of New York, just over a little tree-lined creek from the Audubon Room. Over two days, fifty planetary geologists have been discussing the merits and failings of about twenty different sites on the planet's surface. Each has its champions, each its critics.

Such debates have been at the heart of astrogeology since the 1960s. The choice of landing sites was the most important of the

USGS's contributions to the Apollo program; when it came to landing sites, the biologists whose studies were the *raison d'être* for the *Viking* missions to Mars took the back seat as geologists worried about where their ships might safely come to shore. Geology, with its intense concentration on terrain, comes into its own at such times. If humans go to Mars in person, geologists will choose the landing sites, at least at first.

Some geologists revel in this: Vic Baker sees it as proof of the special nature of the bond between the geologists and their subject, a bond that geophysicists and their like can never have. He points to the dwindling part played by astrogeologists in the science of the 1980s, when no one was landing on anything, and to a possible renaissance in the 2000s, now landings are back on the agenda. Others are less enthusiastic: Peter Schultz at Brown University feels that the adversarial nature of the process of picking sites, in which every geologist argues for his or her own favorite and against the others' candidates, has made the whole field too easily polarized. It has reinforced a tendency that anyone committed to the method of multiple working hypotheses must guard against: that of defending established positions rather than entertaining new ones. "It's wrong to act like lawyers," he says.

There's doubtless some truth in this. But if so, then landing-site selection is still a necessary evil. Missions that land add to our understanding of a planet with data that simply can't be sensed from orbit. There is no way to explore without them. And as they add to the science, they also change its context. They provide us with the opportunity to create real places.

In his classic study *Place and Placelessness,* the Canadian geographer Edward Relph explains how much more there is to a place than the objective facts of its spatial coordinates. "In our everyday lives places are not experienced as independent, clearly defined entities that can be described simply in terms of their location or appearance. Rather [they] are sensed in a chiaroscuro of setting, landscape, ritual, routine, other people, personal experiences, care and concern for home, and in the context of other places." In debates about landing-site selection places are, indeed, seen in the far-from-everyday way Relph describes, possessed of nothing but location and appearance, and this is one of the things that makes such discussions odd. But once a lander has descended to its site, some of Relph's chiaroscuro descends with it. The sight of the surface seen from the surface provides a sense of extension, of setting and landscape. The duration of the mission gives the site an existence in time as well as space, and that allows the development of rituals and routines. The landing site becomes, among the people studying it and building those routines, a shared location for personal experience. It becomes a thing to care about. And this all matters. Though a planetary perspective is a magnificent and enriching thing, places, not planets, are the core of human experience. It is from places that we build our world.

The philosophy of place is one thing. Actually choosing landing sites, though, is something else: a multilayered exercise in frustration, as the drinkers in the Audubon Room can attest. The most obvious frustration is that, contrary to the expectations of the happy-hour rank and file, the workshop has not really whittled away the list of candidate sites. So although there has been a contest of sorts, there is no win-or-lose closure for the contestants. A more profound frustration is that the discussions only served to highlight the limitations of the spacecraft involved; there would be vastly more to do at any one of the possible landing sites than the spacecraft involved is capable of. Worst of all, the meeting has rubbed everybody's nose in how profoundly mysterious Mars really is. In the abstract, that mystery is no bad thing; it's what drives the process of exploration. But when the mystery means that you simply don't have the data you need to make even partially informed choices about where to land your spacecraft it's simply a pain.

Given all this frustration it is just as well that the assembled company cannot know that their labor is largely in vain. The Mars 2001 lander was canceled the following winter. Conceived in response to one failure—the explosion of *Mars Observer* in 1993—it was canceled in the wake of two more, *Mars Climate Orbiter*'s incineration prior to its aerobraking maneuver and the *Mars Polar Lander*'s crash (or "lithobraking maneuver," as the crueler wits had it).

The charter for the original Mars Surveyor program, set in motion after the loss of *Mars Observer,* was that it should send an orbiter and a lander from the Earth to Mars every time the planets' orbits permitted it, which means every twenty-six months or so. The first three orbiters—*Mars Global Surveyor, Mars Climate Orbiter,* and the 01 Orbiter (later christened *Mars Odyssey*)—had a fairly straightforward remit. They were to fly duplicates of *Mars Observer*'s instruments, five on the first one and two each on the second and third. The landers had a more diverse agenda. *Pathfinder** was basically just meant to prove that it was possible to go to Mars for a lot less money than it had cost to send the *Viking*s there. *Mars Polar Lander* had a science payload focused on the frosts and frozen soils of the high latitudes that had been under discussion since the early 1990s.

Not long after *Mars Polar Lander* was approved, though, the Surveyor program underwent a dramatic shift. The reason was the announcement in the summer of 1996 that the oldest of the Martian meteorites, ALH 84001, contained structures that might be fossilized microorganisms. Though within the scientific community this was—and remains—a controversial claim, it certainly rekindled interest in Mars. President Clinton announced his determination that "the American space program will put its full intellectual power and technological prowess behind the search for further life on Mars." But this did not happen. Instead, NASA slightly increased the budgets of the Surveyor program, focused it more on biological issues—which irked some geologists fascinated by Mars as a planet whether or not it has ever borne life—and greatly increased its ambitions. The fourth lander to be launched, in 2003, would now not only carry a rover capable of wandering quite some distance from the lander to retrieve samples, but also a rocket capable of lifting those samples into orbit round Mars. In 2005 another lander/rover/rocket team would repeat the process at another site. A separate spacecraft would gather the samples up from their orbits round Mars in the manner of the evil Blofeld's spacecraft-swallowing spaceship in *You Only Live Twice,* but with kindlier intentions.

**Pathfinder* was not strictly part of the Surveyor program, since it was in the pipeline long before *Mars Observer* was lost. But it was certainly part of the same "faster-better-cheaper" approach to Mars exploration, and retroactively started to be perceived as part of the same enterprise.

That ship would deliver the samples back to Earth where they would land, not in an exquisitely designed extinct Japanese volcano, but in the Utah desert. The Surveyor program was being inflated into the most ambitious set of unmanned space missions ever attempted. And yet it was still to be done on the cheap. Even after the French became partners, promising to provide launch vehicles and the sample-snaffling return-to-earth vessel, the budgets were very tight. And there was still a commitment to launching something at every opportunity.

Originally, the 2001 surface mission (which was to use the same design as *Mars Polar Lander* and to cost considerably less than *Pathfinder*) was meant to test a new rover. With the first sample return mission pushed up to 2003, though, there was no longer enough budget to test the rover it would rely on in 2001. So the 2001 mission lost its rover; the scientific instruments for the 2003 rover—a package called Athena—would be mounted on the lander itself and tested from there. Amid worries that this made the mission embarrassingly uninteresting, a new rover was added to the mission—but only a little *Sojourner* clone. Still the budgetary pressure was intense. Just a week before the Buffalo meeting there had been an attempt to get the "01 Lander"* canceled and its budget shifted to the sample-return project.

One way the 01 Lander might have made its mark was if it were to have had a spectacularly cool site to go to, a target that thrilled the scientists and that would also look pretty exciting to the public. For the geologists gathered in Buffalo, though, finding such a site was easier said than done. Because of the limits of its launch vehicle and the time of year at which it would arrive, the lander could only be targeted at latitudes between 3° N and 12° S—a band that covers only about 13 percent of the planet. It could not land anywhere where the ground sloped too steeply, or was too thickly mantled with dust. It could not land at high altitudes (because its parachutes needed relatively thick air to slow it down); nor could it land in deep depressions (too much atmosphere might interfere with the deploy-

*Tellingly, the mission was never known by any other name than that of the slot on the schedule that it served to fill.

ment of its solar panels). Its legs were short and so it could not land anywhere where there was a significant chance of it disemboweling itself on a rock. The landing site had to be a lot less rocky than *Pathfinder*'s site at the end of Ares Vallis, or *Viking 1*'s a few hundred miles northwest across Chryse, or *Viking 2*'s in Utopia.

Those *Viking* landing sites were themselves the fruit of a complex and tense site-selection process, one that had begun seven years before the spacecrafts' launch and had continued to within days of their landings. Throughout the early 1970s, various committees used the pictures that *Mariner 9* had sent back to decide on possible landing sites; when the *Viking*s went into orbit around Mars in 1976, huge amounts of that work were more or less thrown out. The images sent back by the *Viking* orbiters—to which the landers were attached prior to their final descent—were far more detailed than the *Mariner* images, partly because the cameras were better and partly because there was much less dust in the air. "We were just astounded," according to Mike Carr, who was in charge of the orbiters' cameras (and had fought off repeated attempts to have them either downgraded or removed as a cost-cutting exercise). "It was a mixture of elation and shock." The site chosen for *Viking 1* was revealed as far rougher than had been expected, covered with sharp erosion features and small craters that had been invisible to *Mariner 9*.

Hal Masursky, who was in charge of "certifying" the *Viking* site as safe, set up a massive, highly focused mapping effort more or less overnight. Scores of student volunteers found themselves counting craters; planetary geologists from the USGS and JPL were drafted in to interpret the hitherto unseen landforms. When images of the backup sites were returned from the spacecraft, they seemed as off-putting as the primary. The spacecraft's orbit was changed to scout out new areas. But the pictures weren't the only things keeping Masursky and his team up at night. Some of the largest radio telescopes on earth—including the wire-mesh-lined valley at Arecibo in Puerto Rico, and the seventy-meter dish at Goldstone, California, originally built to listen to *Mariner 4*'s faint whisperings—had bounced radar beams off the Martian surface. The echoes they got back couldn't match the images taken from orbit feature for feature:

The radar beams illuminated thousands of square miles of the surface at a time. But radar could measure the average roughness of these swathes of the surface on scales as fine as four inches. Some people involved in the site selection, notably Carl Sagan, thought the radar measurements were more important than the images sent back by the orbiters. Again and again Sagan pointed out that what a surface looks like at hundred-yard resolution—roughly speaking the best that the *Viking* orbiters could manage—may have nothing to do with what it looks like at one-yard resolution; and one-yard roughness would be quite enough to kill a *Viking* lander.

The radar returns did not just reveal something new about Mars; they highlighted something fundamental about how limited an approach simply looking at a planet is. Orbital images and the wonderful maps produced through their study are both a magnificent scientific achievement and something akin to a deception. The deception stems from our assumptions about what a map is. Everything we know tells us that big pictures come from little pictures, that the truth about something being mapped comes from the bottom up, local knowledge made ever more universal and objective as surveyors link places together. Mapmaking is taking a step back. So it is natural to assume that images of Mars—the airbrush maps and the orbital images and the corrected photomosaics and the MOLA topography and all the other space-age representations—are summaries and generalizations just like all our other maps. But for Mars, the maps are all we have. Everything in them was learned from outside, and everything that could be known is right there on the sheet or screen—they are the result of zooming in, not pulling out. They are not built up out of places that have been experienced, or even described. There is no higher level of detail that can be accessed; a magnifying glass will just show you the dots of the printing process.

Viking 1's controllers "walked" their orbiting spacecraft farther and farther from its original destination while the geologists pored over their images looking for a landing site. Candidate sites that looked far safer than the original turned out to be just as rough in terms of the radar return, and the unseen trumped the seen. Only on 12 July—eight days after the symbolically weighted date for which the first touchdown on another planet had been slated—was a site found that looked smooth to the orbiters' eyes and sounded flat to the radar ears. *Viking 1* landed about five hundred miles northwest of its original target site on 20 July 1976, the seventh anniversary of Neil Armstrong's touchdown on the moon. A little more than six weeks later, *Viking 2* landed at its second backup site in Utopia after the primary in Cydonia and the first backup had proved similarly worrying. It was a close-run thing; analysis of the rocks among which it found itself suggests that *Viking 2* ran an almost one-in-three chance of coming down on a crippling boulder.

Hal Masursky and his colleagues spent weeks working twenty-hour days in order to decide which barely understood rock-strewn wilderness they would throw a billion dollars' worth of lander at. Twenty years later Matt Golombek, the chief scientist for *Mars Pathfinder,* had no such luxury. *Pathfinder,* like all the landers in the Surveyor program, lacked the motors that would be required if it were to put itself into orbit round Mars before landing; it was to go directly from interplanetary space to the planetary surface, a meteorite with a parachute. So there would be no time for deliberation, backups, and second-guessing. For Golombek, choosing and certifying the landing site was the mission's most formidable challenge and it was more or less entirely on his shoulders.

If *Mars Observer* had reached orbit in 1993, as he originally expected it to, Golombek's job would have been much easier. Mike Malin's camera could have been put to work examining potential landing sites in far greater detail than the *Viking* cameras did. But *Mars Observer* died. And though a backup of the MOC instrument was to fly three years later on board *Mars Global Surveyor,* it would arrive at its destination well after *Pathfinder.* So Golombek and the team he gathered around him (notably *Viking* veteran Hank Moore from the USGS and Tim Parker, whose expertise at interpreting pictures of Mars is highly prized at JPL) chose their sites with only the data already in hand. First they thought about where they might go to find interesting geology: Although *Pathfinder* was basically an engineering mission designed to test a new landing system, it did carry instruments that would help geologists work out what sort of rocks it was seeing. Once they had narrowed the good geology sites down to three, they used everything that was known about the short-listed sites to create what Golombek called "the chart from hell," an imposing tabulation of potentials and pitfalls.

Golombek may have had to rely on old data, but he had the advantage of a new type of spacecraft. Rather than lowering itself down onto its landing legs with neat little bursts from its retro-rockets, as the *Viking*s had done, *Pathfinder* would hit the planet wrapped in air bags, bouncing across the surface before opening up like a flower at its final resting place. Because of this robust if inelegant approach, rockiness did not worry *Pathfinder* much—a good thing, considering it was a geology mission to a rocky planet. Rather than avoiding the rockiest sites, as the *Viking*s had tried to do, Golombek was in the business of seeking rocks out, and one of the data sets on the chart from hell told him where to look for them. Infrared measurements from the *Viking* orbiters showed how quickly the surface cooled down at night, and by knowing that bare rock cools slowly and dust cools fast it was possible to estimate from the spectrum how rocky a given stretch of surface might be. This technique worked pretty well, landing *Pathfinder* at a site more or less exactly as rocky as Golombek had intended. "You want a landing site," he crowed to his colleagues, "I deliver!"

As the man who had delivered a landing site most recently, Golombek was one of the first people to speak at the Buffalo workshop where sites for the 01 Lander were to be chosen. His main message was very simple: Safety is everything. It doesn't matter how wonderful the scientific potential of a given site on Mars may be; if the spacecraft runs a risk of not surviving a landing there, it shouldn't be sent there. And there would be an awful lot of Mars where the 01 Lander would not survive. Rather than bouncing to a halt on airbags like *Pathfinder,* the 01 Lander lowered itself down on rockets, exposing a delicate belly to any rock that might stick more than fourteen inches off the ground between its feet. The high rock abundances that had been magnets for *Pathfinder* were reefs to be avoided for the 01 Lander. If its landing site were anything like *Pathfinder*'s, or either of the *Viking*s', it would risk being gutted.

(The *Viking* landers only had an eight-and-a-half-inch clearance, and had the rockiness of their landing sites been known the landings would not have gone ahead.)

It was vital to get high-resolution MOC images from *Mars Global Surveyor* to understand any potential landing site, Golombek said. But this was a problem. There were still comparatively few MOC images available to the people putting forward landing sites. The final mapping phase of the *MGS* mission had only started that February and the data gathered had not yet been released to the world at large—most of the best MOC data had only been seen by the MOC team members. The fact that MOC images were necessary to make a case for a landing site, and no one had enough of them to do so, was one of the many frustrations of the Buffalo meeting. Ken Edgett was able to offer the people attending general rules that he and Malin had come up with for guessing what a MOC image might contain on the basis of *Viking* pictures of the same site. The best of these was that sites that looked smooth in *Viking* pictures looked rough when seen through the eyes of MOC and vice versa. Since most people were proposing sites at least in part on the basis of *Viking* pictures and had, reasonably enough, stayed away from rough-looking bits, applying the smooth-is-rough, rough-is-smooth rule often meant starting again.

Some people proposed sites based on a systematic tallying of relevant images; some proposed sites that new data or new theories suggested were interesting. There was one where the *MGS* spectrometer had suggested there might be hematite; another where the strong magnetic anomalies had been seen. And there were sites based on long-standing interest. Baerbel Lucchitta and colleagues suggested sites in Valles Marineris. Nathalie Cabrol suggested a site in Gusev crater, which over the years has become something like a second home to her and Edmond Grin; in their imaginations they have spent days walking its rim and gazing down over its inner terraces to the lake beds beyond. They have sung its praises to their colleagues as though selling timeshares. It wasn't a very good site for the 01 Lander, too far south and with its interesting features far too far apart, but Cabrol put it forward anyway, to audible groans from some in the crowd. She defended it vigorously almost as a point of principle: Gusev was the right place to go. If it wasn't right for this lander then that just showed that landers should be designed according to the needs of science, rather than science having to be accommodated to the design of landers.

Ken Edgett offered a peculiarly thought-provoking site: the middle of Ganges Chasma, a cul-de-sac canyon northeast of the main trunk of Valles Marineris. Ganges contains one of the intriguing layered deposits seen in the large canyons, a vast table almost as tall as the canyon is deep, possibly made of sediments, possibly, as Lucchitta's colleague Mary Chapman has suggested, created by underwater volcanism at a time when the chasm was an ice-locked lake. But these deposits weren't Edgett's target. He was putting forward a part of the canyon floor that was featureless in both *Viking* and MOC images, that looked smooth in the MOLA altimetry and that infrared measurements suggested was made of relatively coarse, compacted sand. It was as safe a site as you could wish for. It was also terribly dull. There would in all likelihood be no rocks nearby for the lander's little rover to sniff at. Once the lander had analyzed a few scoops of sand and assured itself that it was, indeed, sand, there would have been nothing for it to do except peer at the layered deposits on the horizon. Seeing them side on, Edgett suggested, might reveal important clues to their origin that would always be hidden from orbital eyes.

Edgett has a strong subversive streak. He also has a sensitivity to how hard running a spacecraft is and a sense of how important it is to minimize risks. Sitting behind him in the Buffalo auditorium, drinking with him in the Audubon Room afterward, I never quite satisfied myself as to whether his proposal to put the lander down on the nearest thing to a completely featureless plain he could find was meant as an extreme but plausible response to the mission's constraints or as a mocking *reductio ad absurdum*. Maybe it was both. But it did have one clear advantage that other proposals lacked: It avoided the odd double standard with which some other people seemed to be thinking about their sites.

The uncertainties expected in measurements of a spacecraft's position while in flight mean that you can't say exactly where it is—or where it will land after passing through an unpredictable atmosphere. The target for a lander is thus not a specific point, but an ellipse that takes these uncertainties into account. Sometimes the ellipse is more or less a circle, sometimes it is highly elongated: The landing ellipse for *Mars Polar Lander* was 12.5 miles wide and 125 miles long, reflecting the fact that the spacecraft was grazing the pole of the planet, rather than hitting a site at the equator straight on.

For the 2001 mission the ellipses were roughly circular and about 12.5 miles across—areas the size of large cities. In one sense these ellipses were the landing sites, in that they were the only answer that could be given to the question "where will the spacecraft land." But ask what the spacecraft could do or see and the answer would be very different. Using a rover essentially identical to the 01 Lander's, *Pathfinder* had explored roughly a thousand square feet of the Martian surface. So the area that the lander and rover would actually get to poke and prod was an invisible third of 1 percent of 1 percent of 1 percent of the area defined by the landing ellipse. Anyone who said that the lander would end up "near" this or that interesting feature was basically saying that the lander would be too far from that feature to study it in any detail. Targeting a specific feature on a map or in an orbital image would be pointless—inappropriate place-based thinking in a context that called for a planetary perspective.

Instead, you needed to go to a site that offered something of interest all across the ellipse, a location whose charms were not local. Long after Buffalo, the site chosen for the 01 Lander—just a few months before it was canceled—was just such a site, a part of the area in Terra Meridiani where *MGS* had detected hematite. There was a chance that the spectrometer on the rover and the camera on the lander's arm would be able to get up close and personal with a sample of rock that might once have been on a Martian sea floor. There was even the possibility—incredibly unlikely—that that old seabed might contain fossils. But there were no particular features of interest on the maps; there was, to borrow from Gertrude Stein, no "there there."

Putting Together a Place

"Welcome to the desert of the real."

—Morpheus in *The Matrix*,
written by the Wachowski brothers

No there there—yet. What *Pathfinder* showed, even more than the *Vikings*, was that once you get to a landing site, no matter how generic, studying it even from a distance feels like an exploration. And the extent of that exploration in time and space gives the site something of the subjectivity of a place inhabited.

Unlike *Viking*, *Pathfinder* landed on the day it was meant to: On 4 July 1997 it bounced and rolled to a stop near the mouth of Ares Vallis. Its airbags deflated and a little mechanism pulled them to one side so that the *Sojourner* rover would not have to fight its way across them. The twin lenses of the spacecraft's camera, IMP (Imager for *Mars Pathfinder*) sent their first images of the landscape back to JPL. Presenting them to the waiting journalists, Peter Smith, the principal investigator on IMP, waxed poetic, "The eyes of the camera are our eyes, and in that sense we are all on Mars. We are there together. You might say that the people of Earth are the soul of this robot. So for the first presentation of images, forget about the engineering and scientific aspects. They're very important, but open your imagination to the experience and beauty of the landscape of Mars."

IMP revealed a flat, yellow-brown plain of rubble under a sky of a similar hue. Golombek had expected the site to be rocky and suspected that the rocks would be from a variety of different sources. Field trips to the channeled scablands in eastern Washington had convinced him that if *Pathfinder* went to the region where Ares debouched into the lowlands of Chryse—more or less the same site as the one originally planned for *Viking 1*—its rover might get to sample wandering rocks from hundreds of miles upstream while only rolling a few yards across the surface. Whether the rocks it looked at did indeed come from different places is not quite clear; but the landscape certainly looked like the aftermath of some sort of devastation.

On the horizon there were various knobs and crater walls. Most prominent were a pair of little knobs immediately dubbed Twin Peaks, which were a particular delight to Peter Smith. In devising a logo for IMP he had used the M to form a double hill on the horizon; Martian nature was imitating his art. Tim Parker and Matt Golombek were happy to see the unusual double formation too—it made pinpointing the landing site in the *Viking* images of the area much easier. Once the Twin Peaks had been found, the position was pretty well known; as other sights on the horizon were matched up with less distinctive features in the *Viking* images, they were able to locate the landing site to within a hundred yards. The protoplace had a fixed location.

Closer to IMP's eyes, the different rocks in the jumbled waste quickly took on their own characters and names. This naming process had none of the formality of the system that the IAU uses for craters, valleys, and the like; according to Dan Britt of the University of Tennessee, who coordinated what system there was, the criteria were that names should be easy to remember, that they should not be the names of real people, and that they should not be derogatory. They should also be something that Britt could spell the first time he heard it. And so there was a Half Dome and a Photometry Flats, a Zaphod and an Indiana Jones, a Soufflé and a Pop Tart, a Moe and a Stimpy, and a Scooby Doo and a Barnacle Bill. There was only one serious name. The lander itself, no longer a spacecraft, became a scientific installation: the Carl Sagan Memorial Station. Sagan had died just before the spacecraft's launch.*

*There was precedent for this; the *Viking 1* lander was renamed the Mutch Memorial Station in honor of Thomas Mutch, a planetary geologist from Brown University who played a key role in the *Viking* mission and died in a mountaineering accident in 1980.

The rocks were minutely observed by the IMP team, which grew ever more sophisticated in its ability to wring out as much information as possible from images that had to be taken from a single spot. With the help of the USGS at Flagstaff, Smith and his team built up a digital model of the terrain and then draped the data from the camera over that model, a trick that allowed them to build up stereographic images like those pioneered by Carleton Watkins, the West Coast landscape photographer, a century before. It may sound like a parlor trick, but it helped the scenes make sense, especially to the untutored eye; among other things, it brought out ripples in the middle distance like the ripples in the bedrock of the channeled scablands.

Perhaps the most striking of all the *Pathfinder* images is another of those in which a computer model is used to reorder the data. The image in question looks down on *Pathfinder* from directly overhead, showing the landscape that it inhabited in the finest detail. The farther from *Pathfinder* you go, though, the more distorted everything becomes. Gores of ignorance appear behind rocks the camera can't see over; oblique surfaces begin to smear out. The landscape takes on the distinctive radial blurring of a jump into hyperspace in one of the *Star Wars* movies.

In one of the corridors in the USGS Flagstaff campus, the *Viking* frame in which the *Pathfinder* landing site sits is blown up to wall size, its pixels square and distinct. Over nine pixels in the spot where *Pathfinder* must surely be found (but in which it has never been resolved) sits a contour map of the terrain generated from IMP. And in the center of that block of nine sits the synthetic overhead view of the *Pathfinder* site in all the glory of its rushing-toward-you blurriness. Nothing could better dramatize the contrast in scales between a planet seen from orbit and a place seen from the ground. *Pathfinder* is locked into a single pixel.

A place is not a place, though, if all you can do is look at it; for a place to be a place, you must also be able to move around it, to go away and look back, to see it from more than one angle. And for this the *Pathfinder* team had *Sojourner*, the first rover to be let loose on Mars. *Sojourner*—its full name was *Sojourner Truth*, after an American abolitionist—trundled from rock to rock, pressing its alpha-proton X-ray spectrometer to their surfaces to work out what the chemical elements within were. (Because the spectrometer was not very well calibrated, these measurements were not in the end all that they could have been.) It was *Sojourner,* more than anything else, that turned the landing site in Ares Vallis into a place. She moved through it and saw it from different angles. She discovered things in it that IMP could not see—behind the area called the Rock Garden there was a beautiful little sand dune. Her slow movements brought time to the land, and time is as necessary to a sense of place as space is.

Smith and the IMP team expended a lot of effort trying to capture time with their camera, but to little avail. They revisited spots where they thought that things might change—where the wind might move dust across a rock, for example—but never found anything. All that changed was the sky, and thus the light. Shadows crept around the rocks, clouds came and went above the horizon: In one of the few purely aesthetic touches to the mission, Smith captured a beautiful pale sunset. He thought about trying to capture the Earth, then Mars's morning star, but it wasn't possible: The dust in the sky meant that dawn began three hours before sunrise, making the rising Earth invisible.*

Regardless of the changes *Pathfinder* saw in the sky, the only change on the surface of Mars was the tracks *Sojourner* left in the dust. But this was enough. *Sojourner* put activity into the images and brought change to the changeless surface. Her tracks in the dust

*Since Tennyson writers evoking Mars have often made much of the sight of the Earth and the moon as a double evening or morning star. Whatever symbolic power such a sight might have, though, as pure spectacle it would be distinctly underwhelming. The Earth will never be as bright in the Martian sky as Venus is in the earth's: Venus comes closer to the Earth than the Earth comes to Mars; Venus is closer to the sun than the Earth and thus illuminated more brightly to begin with; and Venus's even white clouds reflect more sunlight than the Earth does. Indeed, for much of the time not only is Venus as seen from the Earth brighter than the Earth as seen from Mars: Venus as seen from Mars is brighter than the Earth as seen from Mars. As for the Earth's moon, it will be less bright as seen from Mars than Mercury is as seen from the Earth. The Earth will have one subtle but unique characteristic in the skies of Mars, though; while the other planets all change in brightness purely due to their orbits, the Earth's brightness will change capriciously, depending on the cloudiness of the illuminated crescent. Even at these distances the Earth's restless activity will be evident: It might be twice as bright one day as the next.

brought time and motion to Mars. They made the landscape a place of purposeful activity, rather than just a site for disembodied study, and that gave it a new drama. The images sent back recall the way that the presence of explorers—and indeed exploiters—became a motif in many of the classic photographs of the American West made in the nineteenth century. Perhaps the most famous is Timothy O'Sullivan's *Sand Dunes, Carson Desert, Nevada*, a document from an 1860s survey undertaken by Clarence King, who later became the first director of the USGS. The photographer's mule train waits between the dunes and his footprints lead up to the vantage point from which the picture is taken. These were not images trying to evoke an idea of the land, like the landscapes painted by the Hudson River School; they were images of engagement with new terrain.

Most images of other planets do not have this quality, but there is one great exception: the pictures taken by the Apollo astronauts. The photographs they took of their landing sites are recognizably images of places where people walk and work and leave telltale traces. Rover tracks defined the moon's dusty hills just as railway tracks defined the Columbia River landscapes of Carleton Watkins in the 1880s. The story they traced over the lunar surface added to the grandeur of the location.* *Sojourner* did something similar in miniature as it trundled around its rock garden.

The Apollo missions came close to defining the notion of a "media event"—something like the Olympics or a royal wedding, a happening people tell each other they must all witness and that binds an audience together for a shared moment. *Pathfinder,* on the other hand, was a new-media event. Even leaving aside the lack of astronauts, the *Pathfinder* mission did not have the focused drama of an Apollo landing—there were no pictures of the surface rushing up at you, no moments that had to be witnessed. But it had a drawn-out fascination that made it perfect for the World Wide Web, dull to watch in real time but perfect to check in on now and then. The mis-

*The fact that the Apollo photographs, like O'Sullivan's and Watkins's, are works of art as well as technical records is brilliantly demonstrated by the San Francisco–based landscape photographer Michael Light in his 1999 book and exhibition *Full Moon,* which used beautifully remastered digital versions of a hundred or so photographs from the vast Apollo archive.

sion's Web site got something like half a billion hits in July 1997, which was the most any single event, or thing, or place had ever seen. It was hailed as a defining moment for the Web as a purveyor of news—it was to the new medium, suggested the *New York Times*, what President Kennedy's assassination had been to television news or the Gulf War to CNN. The fact that *Pathfinder*'s main achievement—landing on Mars—was a reprise of something done decades earlier could have made the mission seem old hat, but the Web dimension made it seem utterly contemporary. At a time when people were discovering they could see all sorts of distant parts of the Earth through little Web cams, the ability to see somewhere beyond the Earth in just the same way was at once compellingly different and excitingly the same; it made that world part of this one. Even *Pathfinder*'s problems were the problems of the day, with various modem difficulties and an unreasonable need to reboot all the time.

In 1969 NASA administrator Thomas Paine said the moon landing was "the triumph of the squares . . . run by squares, for squares"; *Pathfinder* was run by nerds, for nerds, at a time when a nerdy love of the technological was becoming cool. The youthfulness of the team at JPL, the cartoon names for the rocks, the fact that the whole thing started over the Fourth of July weekend, and the toylike nature of the little rover all came together to make Mars feel like fun. Every place is a place of a certain sort. What sort of place was the *Pathfinder* landing site? It was a playground.

To navigate around the playground a team from NASA Ames produced a geometrical model of the terrain's ups and downs and draped IMP images over it to provide a three-dimensional rendition of the ground around the lander. It was a process very similar to that in which data from orbital cameras is draped over a control net, and they called the product Marsmap. *Sojourner*'s controllers used the Marsmap virtual reality to plan the little robot's moves, all of which had to be preprogrammed, since the time it took light to cross from Earth to Mars made real-time control impossible. (The original Mars mappers, Pat Bridges and Jay Inge at the USGS, had done something similar for the *Viking* landers, modeling the landscapes seen in their cameras to allow precise planning of the landers' arm movements. Given the technology of the day, though, they had used polystyrene for their modeling, sculpting their materials into the shapes of rocks around the landers with hot knives and hairdryers.) As the mission went on, the Marsmap became more complex: When *Sojourner* took pictures they would be added to the model, not as fleshed-out parts of the three-dimensional framework, but as two-dimensional surfaces sitting among the rocks like weird *trompe l'oeil* billboards. The virtual reality thus took on the strange feeling of an exploded comic book, different frames here and there. It developed a beguiling cut-and-paste feel, growing at once ever more stylized and ever more useful.

The artist David Hockney holds that a simple photograph can never capture a sense of a place, because it denies the place any sense of time and the observer any sense of movement; a painting, on the other hand, reassures the eye through brushstroke and gesture that its relationship with the landscape is one that has endured, at least for a while, and its distortions of perspective introduce the idea of movement. Trying to capture that same sense of time-in-place with the tools of photography led Hockney to experiment with collages of lots of different frames—what he calls "joinups" and planetary scientists call photomosaics. The fact that the frames are separate and taken in a sequence gives Hockney's assemblages an undeniable sense of time; the fact that they are often taken from different vantage points gives a sense of movement, too, a sense of walking around a place and seeing all its angles at once. When Peter Smith and I met at a Mars conference in Paris in early 1999, shortly after the French role in the proposed sample-return missions was announced, he urged me to head off and visit the Hockney retrospective at the Pompidou Center. It was striking to see the effects that most scientific image processing tries to get rid of in the name of objectivity—the changes in perspective and scale, the different shadows at different times—not just included but celebrated. By refusing to be a single thing, Hockney's distorted composites give more of a sense of place than any single seamless frame ever could, breaking the constraints of perspective to spread the viewer around—and through—the scene.

The Marsmap virtual reality system, it seems to me, echoes some of what Hockney is trying to do. Its strange discovered perspectives, its billboards, its admission that some parts of the place being explored have yet to be seen, end up saying more than perfect pictures in perfect perspective ever could. This is the future of the mapping of Mars: The building up of virtual realities in which all sorts of different data sets with different flaws and omissions are layered in imaginary spaces, data sets that complement and enrich each other. Like earthly geography, the geography of Mars will move from the one-representation-at-a-time world of the map to the multifaceted universe of the geographical information system, or GIS. It's a move as fitting as it is inevitable; after all, the Mars that is studied on Earth has been a set of digital files since the first all-digital camera took the first pictures of it back on *Mariner 4.* Ken Tanaka at the USGS at Flagstaff is overseeing the development of the delightfully named Pigwad (Planetary Interactive GIS-on-the-Web Analyzable Database), a system that aligns data from mosaics, geological maps, MOC, and MOLA. Soon resources like this will be the primary tools in planetary geology, allowing researchers to visualize complex relationships between, say, ground ice, geological units, and topography in any way they want to.

New computer representations, like the sliding colored contours used to look at MOLA data, will change the way we perceive space and distance in this new virtual Mars rather as similar technologies are now doing on Earth. And moving from solid maps to dataspaces and virtual realities will help scientists fulfill their desire to immerse themselves as completely as possible in data from Mars. In 1976 the best technology Carl Sagan could come up with for experiencing the *Viking* lander sites as places was to make a large print of a *Viking* lander panorama, tape it into a cylinder and sit with it wrapped round his head. The successors to Marsmap and Pigwad will make much richer immersive experiences possible, both at the whole-planet level and at the level of specific sites. They will provide data from Mars with a new interiority; though in the real world Mars will still be seen entirely from the outside, in the virtual world it will be experienced from within.

New technologies will also change the perception of time. While a map has to freeze its representation in an eternal now, in a database time is just one more of the ways in which geographical data can be tagged and displayed. Processes such as the growth of a dust storm or the retreat of a polar cap can be dissected day by day. The only limit is the amount of data. At the moment, that limit is quite severe. The amount of data that can be sent back from a mission is highly constrained and so many moments that might be observed are not. Just as parts of the *Pathfinder* site were not seen, leaving odd blanks in the Marsmap virtual reality, so much of the time that *Pathfinder* passed there went unobserved and unreported. *Pathfinder* was not a continually observing presence; it sent back only four complete panoramas of the place it was sitting in, separated by days or weeks. The amount it missed was demonstrated by the fact that it only caught the fuzziest few glimpses of the dust devils that, other data suggest, were dancing past it throughout its stay. *Pathfinder*'s meteorology instruments suggest that at least one dust devil passed right over it, its shadow felt as a drop of voltage in the spacecraft's solar cells.

The problem is bandwidth. At the moment, each mission to Mars takes its own system for radioing data back to the great antenna at Goldstone and its siblings in Spain and Australia, and capable though those systems are, they represent a terrible bottleneck. In the future, though, this may change. There are plans to set up a system of communication satellites round Mars to ferry data back to Earth using protocols adapted from those of the Internet. The near-term reason for doing this is to remove the need for every spacecraft to carry Mars-to-Earth communications systems, thus allowing missions to the surface to be smaller and lighter; eventually such satellites might also provide guidance for rovers on the Martian surface, as GPS satellites do on Earth. But dedicated communication satellites could also increase the total data traffic between the planets. If they were eventually to use lasers, rather than radio, to link Mars to Earth they might increase the data rate by orders of magnitude, making real-time streaming video a possibility for future missions.

In the long run, this infrastructure may come to be as important in transforming the human experience of Mars as the data that come back through it. Infrastructure has a powerful symbolism: Look at the number of times that the coming of the railways is used in Westerns to mark the passing of an era. The Internet on Earth has taken on a symbolic importance independent of the network's real capabilities. The rhetoric of connection and communication plays an ever-greater part in our discourse, just as the rhetoric of speed, power, and precision did in the wake of the introduction of the railway. An Internet that extended to Mars would be an important statement in and of itself. And its product—an increasingly rich corpus of data from which ever-more-powerful computers could produce ever-better Martian virtual realities—could bring Mars into the everyday lives of those with an interest. Pay a premium to send e-mail to a like-minded friend through a server in orbit around Mars; choose Olympus Mons as a backdrop for your video conference; fly your flight simulator along Echus Chasma; check in with the rover trundling around Hale in search of a lake bed once a day; download the whistle of a Martian wind as the ring tone for your mobile phone.

On Earth, information technologies are often criticized for reducing the importance of a sense of place, of somehow eroding geographical reality. The same charges were laid at the door of the railway revolution in the nineteenth century. There was some truth in the charge then and there is some truth now; railways, and later aircraft, clearly changed the way the world was experienced, and the Internet will do the same. But there is a difference between deforming our sense of place and demolishing it. And in the special case of Mars, an empty planet almost devoid of places, the fear will not only be false—its reverse will be true. Rich near-real-time data, a poor if improving substitute for "being there" on Earth, will be a great improvement over the status quo on Mars.

The lesson of *Pathfinder* is that the more accessible small parts of the surface of Mars become in cyberspace, the more placelike they will come to feel. Places need space to exist in, and time to change, and communities to give them meaning. The illusory spaces, asynchronous times, and real communities of the virtual world provide for those needs, in their way. And it is in those spaces, times and communities that, over the next years and decades, Mars will become more and more of a place.

The Underground

> The ultimate scientific study of Mars will be realized only with the coming of man—man who can conduct seismic and electromagnetic sounding surveys; who can launch balloons, drive rovers, establish geologic field relations, select rock samples and dissect them under the microscope; who can track clouds and witness other meteorological transients; who can drill for permafrost, examine core tubes, and insert heat-flow probes; and who, with his inimitable capacity for application of scientific insight and methodology, can pursue the quest for indigenous life-forms and perhaps discover the fossilized remains of an earlier biosphere.
>
> —Benton Clark,
> "The *Viking* Results—The Case for Man on Mars"

New visualization technologies and higher data rates will change the way we see the surface of Mars. But many scientists are eager for something less superficial. Just as European explorers moved from the sea coasts to the continental interiors in their second great age of exploration, so Martian explorers, too, are looking beyond what can be seen from their ships. They are interested in the planet's depths. Much of what matters most about Mars is clearly beneath its surface and invisible to our eyes: lost terrains now covered in sediment, buried lenses of ice, aquifers (if any), magma chambers. And, most important, if also most speculative, life.

The case for life deep below the Martian surface was first made in a paper published in *Icarus*, the leading planetary science journal, in 1992. The authors were Penny Boston, Mikhail Ivanov, and Chris McKay. Ivanov is an expert on microbial life below the surface of the Earth; Boston and McKay are experts on Mars. More than experts: believers. They met in the 1970s, when they were both in a gifted-students program at Florida Atlantic University. From there they more or less coincidentally both moved on to Boulder, Colorado, which is where they were in 1976, the summer of the *Viking* landings. Neither was a planetary scientist—Boston was a biologist, McKay was moving toward astrophysics—but they and a handful of others in their circle became completely fascinated by the new planet that *Viking* was revealing to them. They devoured the technical papers written by the *Viking* team and started to ask themselves what they could do to take things further. Mars became a consuming passion. As scientists they yearned to understand it; as dreamers and adventurers they yearned to go there.

"I've never been involved in another group quite like it," says Boston. "The magical thing was that everyone's personalities dovetailed. It wasn't the relationship of one person to the group that was important—it was the different relationships we had with each other." Boston and Carol Stoker, who shared an office with McKay, gave the group huge energy. Tom Meyer, Stoker's boyfriend, who had moved to Boulder after a few years working on the wonderfully science-fictional 1970s idea of mining sea floor manganese nodules, provided real experience of opening up a new world through technology. (As a child, he had developed a system for relaying recordings through his hometown's fire alarm wiring so that, with the aid of a bizarre coxcomb antenna attached to military-surplus aviator headphones, he could listen to classical music as he cycled around delivering papers. To Tom Meyer, this was a fairly obvious thing to do.) Boston's partner Steve Welch was a great hands-on technician, the sort of person whom anyone doing experiments wants around. Carter Emmart, the youngest of the bunch, used his artistic skills to help everyone visualize what they were talking about. And McKay provided, among other things, an essential optimism. "Chris's main function for me was not getting discouraged by discouraging things," Boston told me a few years ago. "I can get fairly blue—'No one cares, no one wants to do this'—and he is fairly impervious to that. He's a very complicated man—there's a lot more to him than you see on the surface—but not a moody man." McKay was not exactly the leader, but he was to some extent the focus. He was the one undergraduates would come up to and say, "You're the Mars Guy, right?" Being exceedingly tall and thin helped; so did having a mind that was sharp and quick even by the standards of an extremely intense, intelligent group; so did genuine charisma.

They ran seminars; they imagined new ways of making use of Martian resources; they grew radishes in jars containing carbon dioxide at near Martian pressures. They dreamed with a passion of exploring Mars themselves and despaired—inasmuch as McKay's presence would permit—as they watched NASA's shuttle-driven retrenchment. They went on road trips to meetings where Mars would be discussed and sought out like-minded dreamers, people who saw the self-evident truth that Mars was the most important thing, the most exciting thing—the next thing. And they still found time for naked whitewater rafting. A like-minded journalist, Leonard David, called them the Mars Underground; the name stuck.

In 1981 the Underground gathered their network of sympathizers for a conference to marshal the arguments for sending people to Mars. Benton Clark, who had worked on *Viking* at Martin Marietta, the company that built the lander, had written a paper in 1978 called "After *Viking*—The Case for Man on Mars" that had inspired them; shortening his title and correcting its sexual politics, they called their conference the Case for Mars. That meeting was the first in what turned out to be a triennial series. The meetings often had a strange, fringe-y feeling (McKay, in particular, insisted that no one be excluded as a flake) but that was part of their charm. Where else could you see Hal Masursky slowly and carefully study computer enhancements of the "Face on Mars" before delivering his damning verdict: a drawn-out "Naah" spoken with the authority of a man who had looked at Mars as much, if not as meticulously, as anyone alive.

The case they assembled was this. Mars was worth studying because it was much more similar to the Earth than any other planet and had in the past been more similar still. In particular, its rocks might well contain evidence of past life. Robotic explorers could undoubtedly tell us much about the planet. But those who said that everything worth doing could be done by robots—a fairly common view among space scientists, who resent the far greater budgets with which the manned program produces far fewer results—were wrong. The evidence for life might be subtle and hard to parse. It would require real fieldwork and real laboratory skills on site to put it together. And to do it right you would need humans.

The same things that make Mars interesting also make its human exploration feasible—it has an environment that can supply things explorers would need. They could grow plants there in greenhouses irrigated with melted ground ice. They could make rocket fuel from the atmosphere. Life was thus the beginning of the case, because life on Mars in the past was what most needed studying, and the end of the case—because those studies required that we learn how to live on Mars ourselves. Across billions of years and hundreds of millions of miles, life was calling out to life.

Thomas Paine, who had run NASA from 1968 to 1970, was convinced to give a keynote address at the second of the Case for Mars meetings: The man who had celebrated the "squares" of Apollo was delighted by the longhairs of Boulder and their dreams of living off the Martian land. The Underground was invited to Washington to present its case. Meyer remembers overhearing the head of manned space flight lean over and whisper to a colleague, "I had no idea this was going on. We've got to look into this."

NASA did look into it. Paine wrote a much-discussed report, "Pioneering the Space Frontier," that called for the exploration and colonization of Mars, among other things. It attracted the interest of the White House. In 1989, on the twentieth anniversary of the first Apollo landing, George Bush the Elder called for a return to the moon and an outpost on Mars. But the president's enthusiasm got little further than Spiro Agnew's had twenty years before. NASA came up with a spectacularly complicated and ornate plan that included more or less everything its engineers had ever wanted to do and was perhaps ten times costlier than it could have been, which left the idea—for which the president had done little to prepare Congress—dead in the water.

The Underground was not just about proselytizing, though; it was also about science and technology. Stoker became an expert on what rovers can and cannot do on planetary surfaces: She was one of the leaders of the team that produced the virtual reality Marsmap system for *Sojourner*. Boston threw herself into studying the complex interactions within biospheres in order to make sense of what biology could do on a planetary scale; she was one of the people who got the American Geophysical Union to take its first serious look at Jim Lovelock's Gaia theory, which claims that the planet Earth and its biosphere together form a self-regulating system. And McKay started to study life in the seemingly sterile dry valleys of Antarctica, where the temperature is below freezing for almost all of the year and the precipitation is dauntingly close to zero.

It was Imré Friedmann, a microbiologist from Florida State University, who first suggested how life might persist in such a desolate place. In the 1960s Friedmann's search for life in the Negev desert had revealed lichen growing inside sedimentary rocks, staying close enough to the surface for a little light to get through to them, but keeping enough rock between them and the outside world to stop any moisture from getting out. He thought similar "cryptoendoliths" might be found in Antarctica, but he couldn't get a grant to go and find out. However, a friend of his, Wolf Vishniac, had a NASA grant to go to the dry valleys as part of his work on the *Viking* program; they were already recognized as arguably the most Mars-like places on the Earth in their frigid aridity. Friedmann asked Vishniac to pick him up some random sandstones so that he could look for lichen within them. Vishniac died on the trip; hiking into the Asgard Mountains to see whether any microbes had grown on nutrient samples he had left there, he stumbled and fell. But when his effects were shipped home, his wife found some stones in them labeled "for Imré." She sent them on; beneath their crust there was a layer of life. Friedmann published and the world took note. Grants were no longer an insurmountable problem.

By 1980, Friedmann and his wife had made a number of trips to Antarctica to further their research, and all sorts of people were eager to come along. Rather to his surprise, one of the few Friedmann ended up taking was a young physicist from Boulder with no real biological training and no field experience. "I just thought, 'This young man is serious,'" says Friedmann, now semiretired, in a voice that suggests there is no greater attribute under the sun than seriousness. With a chuckle, he adds, "I have a good nose. Chris McKay's the calmest and most balanced man I have ever met. He is also very, very intelligent. I would, without exaggeration, call him a genius." McKay quickly became a key member of Friedmann's team, cooking up new ways to measure the "nanoclimates" that allowed the microbes to keep tiny amounts of water liquid within rocks despite what was going on outside. He also became part of the human glue that holds such ventures together, remaining pleasant and sharp in conditions that can make people alarmingly unstable.

Establishing a reputation in Antarctica led to invitations to study other extreme environments. In Siberia, Chris McKay found microbes that had been frozen for three million years, but that could still reproduce when thawed out. In the Gobi Desert he found microbial life all too present when dysentery-bearing dumplings struck down his whole expedition. In the far north of Canada he found springs that welled up through permafrost a half-mile deep. Everywhere McKay found liquid water and a supply of nutrients, he found life. Only in Chile's Atacama Desert was there no life to be measured. Bacteria fell to the surface from the sky (as they do all over the world) but failed to flourish. And the Atacama was the driest place McKay had been—no precipitation in four years. "It gives us our first data point on the other side of the line: too dry for life," says McKay.

The Martian surface looks like one great big Atacama, except worse. It is not only arid and frozen but bathed in harsh ultraviolet radiation; the Martian atmosphere, lacking oxygen, necessarily lacks an ozone layer. The *Viking* landers had discovered that the regolith appeared to be laced with highly reactive chemicals, which no one had expected and which led to initially confusing results. Experiments trying to detect life by seeing if nutrients were broken down by something in the soil scored a big yes, which led to great excitement. But these reactions quickly tapered off, whereas metabolic reactions would be expected to persist; and they produced gases life would not be expected to produce (oxygen was given off by the samples being incubated, which in the absence of sunlight didn't seem a likely sign of life). What's more, studies of the soil itself showed that it contained no organic compounds. So the soil reactions were agreed by most to be chemical, rather than biochemical. There were dissenters—most notably Gil Levin, who still claims his experiment on the landers found life—but the consensus opinion was and remains that the soil was laced with peroxides produced by the high levels of ultraviolet light* and utterly without life.

A lifeless surface was assumed to be the same as a lifeless planet. In the late 1980s, though, that assumption started to come under renewed scrutiny. The floors of the Earth's oceans had long been thought more or less lifeless; there was obviously water there, but there seemed to be no obvious source of nutrients. With no sunlight there couldn't be any photosynthesis and photosynthesis sits at the bottom of almost all the food webs that biologists study. But just a few months after the *Viking* landings the submersible *Alvin* discovered a profusion of life around volcanic vents in the Galápagos rift, a mile and a half below the surface. Since then many such systems have been found, containing bacteria living under immense pressure and appreciably above the normal boiling point of water.

In the 1980s the SNC meteorites—and in particular the youngest of them, the Shergotty meteorite—suggested that there might still be volcanic activity going on within the Martian crust. If it was close to the surface and near the poles, that volcanism would surely melt ground ice and cause liquid water to circulate. So Mars might well have had hydrothermal systems of its own in the past and might still have them today. If there was life clustered around the lightless hydrothermal vents in the depths of the Earth's ocean, why should there not be life in the hydrothermal systems deep inside the Martian crust?

One answer is that the Earth's hydrothermal ecologies are not independent; they rely on imports from other parts of the Earth's biosphere. Life requires a supply of chemicals willing to give up electrons and a supply of chemicals willing to take those electrons up. At the surface of the Earth the giving of electrons is mostly done by organic carbon compounds—food, if you want to get technical about it—and the taking up is done mostly by oxygen produced by plants. That is why the loss of electrons is called oxidation.

*In Geoffrey Landis's *Mars Crossing*, a novel that tries to deal with a near-future mission to Mars as realistically as possible, the explorers go blond and get itchy as traces of peroxide-laced dust build up in their living quarters.

The reverse process is called reduction (because it reduces oxide ores to pure metals) and the two processes have to happen in concert: If an oxidizer is to gain electrons, a reducer must give them up. Although there's only a tiny amount of photosynthesis at deep ocean vents,* there is a steady supply of oxygen-bearing compounds drifting down from the sunlit shallows above, and they are what the vast majority of the creatures around the vents use to oxidize their food.

On Mars there's no supply of oxygen at the surface to seep into the depths. However, though oxygen is a particularly powerful oxidizer, it's not the only one life makes use of. Some earthly bacteria oxidize hydrogen with carbon dioxide to make methane (or acetate) and water; others reduce water with carbon monoxide to make methane and carbon dioxide. There are bacteria that use hydrogen to reduce sulfates, or sulfur itself, or electron-hungry forms of iron. These reactions don't liberate anything like as much energy as reactions between organic matter and oxygen. But they can still be made to work. If there is a constant supply of some fairly simple chemical feedstuffs to work with, life can make do without any oxygen or organic matter to feed on. And hydrothermal systems—mixtures of groundwater and volcanic gases—are often rich in the highly reduced chemicals needed.

In their 1992 *Icarus* paper, Boston, Ivanov, and McKay laid out the argument that life of this sort might be possible in aquifers deep below the Martian surface. In the same year the famously controversial astrophysicist Thomas Gold published a paper that, while based on a very different chain of reasoning, brought similar ideas about the Earth to a wider audience. Gold has long believed that the Earth and the other rocky planets contain vast primordial hydrocarbon reserves, and that it is the upwelling of these hydrocarbons, rather than the burial of organic matter from the surface, which provides our reserves of oil, gas, and coal, an idea summarily rejected by most geologists. A corollary to this idea is that there is a vast number of microbes feeding off these hydrocarbons in the

*Some deep-sea bacteria apparently photosynthesize using the light given off by their phosphorescent neighbors, which is remarkable but not very significant.

depths of the crust: Gold's paper called them "the deep hot biosphere," and the term started to slip into scientific usage, even though Gold's explanation of the processes that might feed such a biosphere remains very much a minority view.

Soon afterward, an outpost of the deep, hot biosphere was apparently discovered inside the Columbia River basalts in central and eastern Washington State. Deep below the surface, warm water was reacting with minerals in the basalt to release hydrogen, and the hydrogen was apparently being used by bacteria as a source of energy; they caused it to reduce carbon dioxide, thus producing methane. Microorganisms had been found at such depths before—Ivanov was an expert on the subject—but normally in sedimentary rocks and oil fields. Here they seemed to be prospering in an aquifer within igneous rock just like that which makes up the surface layers of Mars.* McKay came across the data in a poster session at a conference and says he's never been more surprised by anything in his life. The idea of some sort of deep, hot biosphere on Mars started to feel a lot more real.

In 1994, Boston and McKay decided to go and have a closer look at the life below. Probably the first spelunkers ever funded by NASA, they and their friend Larry Lemke descended into the vast Lechuguilla cave in New Mexico to look for new forms of microbial life. Unlike most large caves, which are created by carbonate rocks dissolving in water, Lechuguilla was carved by diluted sulfuric acid and might provide intriguing niches for sulfur-metabolizing microbes. The first trip was, says Boston, an awful experience; unprepared for the rigors of caving, by the end her objective "was just getting out alive." McKay thinks it was the toughest field trip he's ever done—which is saying quite a lot. He also found that caves hold little interest for him from a professional point of view—his role in fieldwork is to monitor nanoclimates that persist in spite of changes in the environment outside, and, in general, cave environments are pretty stable—and so after that first expedition he saw little reason to repeat the experience.

*The Columbia River basalts are also the basement rock beneath much of the channeled scablands, which makes them doubly Martian.

Boston, on the other hand, came away fascinated, and now focuses a great deal of her research on the weird and intimate relationships between biology and mineralogy to be found in caves, studying rocks that have life running through their veins and living slimes on the brink of turning to stone. The science is fascinating, but there's something more personal to it than that. McKay talks of finding a certain oneness with the world in the seemingly lifeless wastes and solitary places of Antarctica. That's not why he goes there—and he says he can find it in his office at Ames on a quiet Saturday afternoon, if he's lucky—but it definitely counts as one of the perks of the job. Boston gets closest to that feeling in the closeness of her caves, and that's one of the reasons she returns to their sometimes unbreathable atmospheres, their dripping bacterial "snottites," and their battery acid streams again and again. There is something about life at the edge of where life is possible that makes everything, living and nonliving, seem valuable.

In the early summer of 1996, one of the discoverers of the Columbia River basalt bacteria gave a talk at NASA's Johnson Space Center and provided some samples of the bacteria-bearing rock for the scientists there to study. A summer intern was put in charge of making electron microscope images of them. The images showed what appeared to be small bacteria (small, that is, even by bacterial standards) nestled in tiny cracks in the rock. When the intern showed Polaroids of the images to her boss, David McKay—an expert on meteorites and the moon's regolith, and no relation to Chris—he smiled and passed her a set of pictures that seemed remarkably similar. They were electron micrographs of formations inside a Martian meteorite called ALH 84001, pictures that six weeks later would be on the front page of half the newspapers in the world. McKay and his colleague Everett Gibson had found what seemed to them to be signs of fossilized life in the meteorite a year before, but they had been playing their cards very close to their chests, letting only a few colleagues into the secret as they tried to muster as much support as they could through further studies.

ALH 84001 is the oldest of the Martian meteorites that have been thrown to Earth and recognized for what they are; radioisotope analysis dates it to the very beginning of Martian history, four-and-a-half billion years ago. The carbonates within it are younger, perhaps as much as a billion years younger, but they are still old enough to date from Mars's putative warm, wet phase, which was helpful to the argument that they might contain—or be—signs of life. The rock was kicked off Mars by an impact about sixteen million years ago and landed in Antarctica about thirteen thousand years ago. It was found in 1984 as part of a program to collect Antarctic meteorites and identified as Martian in 1993.

By 1996 David McKay and his colleagues in the Johnson team had found various possible signs of life in their rock. There were small hydrocarbon molecules called PAHs, which might be produced by the breakdown of organic matter. There were carbonates in which the ratio between the two stable isotopes of carbon—carbon-12 and carbon-13—was oddly skewed; such carbon-isotope ratios are often associated with living beings. Some of the carbonates were in shapes that looked like tiny fossilized bacteria and one of them, "the worm," appeared to be segmented. And there were very regular little grains of the iron-bearing mineral magnetite. Some earthly bacteria produce magnetite grains of a very particular size and shape, and the Martian magnetite grains looked just like them. On Earth, according to experts on biological magnetite, no inorganic processes produce magnetite grains with this same crystal structure.

The Johnson team's paper was accepted by the journal *Science* in July. Knowing that its publication in August would set off a storm of interest, David McKay and his team headed off to Washington to brief NASA administrator Dan Goldin. Goldin gave them a thorough grilling and then went to brief the White House. The micrographs were not exactly what Chief of Staff Leon Panetta had been expecting as evidence of life on Mars—after all, this was the summer when according to the movie *Independence Day* alien life was the sort of thing that arrived in spaceships the size of cities and bent on blowing up Panetta's office—but Goldin convinced him that the little worm was a big thing. President Clinton was briefed, as was

Vice President Gore. A few days later, Clinton announced the discovery to the world.*

ALH 84001—Penny Boston calls it "Big Alh"—reshaped the Mars program and it refocused a great deal of NASA's science. It is, in part, due to ALH 84001 that NASA has turned itself heavily toward what is now known as "astrobiology." For all its effects, though, Chris McKay and many others were largely unconvinced by ALH 84001. Carbonate blobs are carbonate blobs, and some experts were saying that these particular blobs could only have been created at 1,300°F, far too high a temperature for any sort of life. Microscope images are notoriously open to interpretation and who knows what processes might have influenced carbon isotope ratios billions of years ago on an alien planet. PAHs are found in all sorts of rocks, including meteorites that have never been near Mars. But the magnetites were interesting. Chris McKay and others at Ames quickly organized a two-day seminar on the subject. One of the speakers was the world's greatest expert on biological magnetism, Joe Kirschvink of Caltech.

In the early 1970s Kirschvink was a student at Caltech at the time that Gene Shoemaker, then head of the geology department, was setting up a lab to look at the magnetic fields trapped in the Moenkopi sandstones of the Colorado Plateau. Shoemaker had an exquisitely sensitive magnetometer built for his work, and over the following twenty years Kirschvink used it for a wide range of studies, looking for biological magnetite in fossils, in racing pigeons, and in human brains. Kirschvink has a broad mind, an argumentative Caltech stance and a tendency to champion provocative, dramatic theories. He's one of the leading advocates of the snowball Earth theory; he's recently reapplied the idea of true polar wander—the notion that Peter Schultz used to try to explain layered deposits all over Mars as the fossils of bygone polar caps, and oddly enough another idea originally thought up by Thomas Gold—to the history of the Earth, arguing that there's evidence for a massive concerted flopping-over of all the lithospheric plates at the begin-

*Well-connected Mars Underground fellow traveler Leonard David got the story a day early.

ning of the Cambrian period 540 million years ago. A man with a taste for the provocative and a profound belief in the significance of biological magnetite, Kirschvink found the evidence of biology in ALH 84001 incredibly exciting. (He has since gone on to work closely with the Johnson team.) Chris McKay left the seminar still not convinced, but even more intrigued by the magnetite, partly because it might fit into a favorite speculation of his own.

Some bacteria—called aerobes—need oxygen, some—the anaerobes—never touch the stuff. Most of the bacteria that produce grains of magnetite are picky in-betweeners. They tend to live in sediments where oxygen levels are low but not too low; the chains of magnetite that run along their bodies keep them oriented along the lines of the Earth's magnetic field and thus help them move up and down in the sediment to the oxygen level that suits them best.* This approach, though, presupposes that there is some oxygen around in the first place. On the Earth, this has not always been true. It is not yet clear when the photosynthetic production of oxygen evolved, but only about 2.3 billion years ago did significant amounts of oxygen start to make it into the atmosphere.

When scientists talk about the possibility of life on Mars early in its history, they normally assume that it would have been purely anaerobic, with little if any photosynthesis. The history of life on Mars would have been a billion years of bugs that went extinct when the climate got worse. A wrinkle on this argument is that the advent of photosynthesis may actually have killed them. James Kasting of Pennsylvania State University, an expert on models of the early climate on both the Earth and Mars, thinks that the only greenhouse gas powerful enough to have given Mars a warm and wet early period would have been methane, and that that methane would probably have had to be biologically produced. Organisms living in a methane greenhouse would be very ill advised to start pumping out oxygen, since it would react with the methane and deprive them of their warmth. On the Earth, according to Joe Kirschvink, the appearance of oxygen 2.3 billion years ago seems to

*For a little thing like a bacterium, the twisting force exerted by the Earth's magnetic field is much stronger than the force of gravity.

have triggered one of the snowball-Earth events that covered the whole globe with ice, possibly by removing methane from the atmosphere. On Mars the effect would have been even more dramatic and probably terminal.

The argument that Penny Boston and Chris McKay had put forward with Ivanov was that whatever had happened on the surface, survivors might persist in deep hydrothermal systems to this day. But McKay has since become intrigued by another possibility: that life on Mars might have moved much more quickly than life on Earth. Maybe life on Mars had started to pump out oxygen very early on. Since Mars was less volcanic than the Earth, the supply of reduced chemicals available to react with that oxygen would have been much less copious, and free oxygen might have built up at the surface where it could be used to power far more energetic metabolisms than those available to anaerobes living off rocks. Mars might have had oxygen in its atmosphere long before the Earth did and might have evolved more complex organisms than simple bacteria. Rather than being a sickly weakling, failing to thrive and never getting beyond infancy, it might have been a James Dean of a planet, living fast before dying young. Its rusted face might be the legacy of that life. Maybe it wasn't oxidized by ultraviolet radiation. Maybe it was oxidized by oxygen produced by photosynthesis. This is obviously hard to reconcile with Kasting's notion of a methane greenhouse, but as both of them would quickly point out, that's why we need more data.

Biological magnetite fits into McKay's speculation in an intriguing way. Kirschvink suspects that on the Earth the original evolutionary pressure that drove bacteria to produce magnetite was linked to the chemical problems faced when oxygen becomes abundant. The way magnetite is used for navigation only makes sense in a world with oxygen gradients in its sediments. So if Martian bacteria were producing magnetite in large quantities more than three and a half billion years ago for either of the reasons that earthly bacteria produce it, it would seem likely that there was oxygen around.

Chris McKay is still not convinced that the magnetite in ALH 84001 is evidence of life, but he's leaning more toward the idea than he did originally. One thing that has swayed him is that his old friend Imré Friedmann thinks he has found magnetite in the meteorite that is actually arranged in chains, as the magnetite crystals in earthly bacteria are. That said, there are others who find the evidence for chains—and indeed the whole argument that the presence of this sort of magnetite necessarily implies that there were bacteria around to make it—poor. And there is no clear reason to believe that the two sides will ever really come to an agreement until there is more evidence to go on. Probably the most quoted thing that Carl Sagan ever said (bearing in mind that he used to insist that he never actually said "billions and billions") was that "Extraordinary claims require extraordinary evidence." On its own, ALH 84001 may simply not be quite extraordinary enough.* But it has certainly continued the recasting of expectations about where Martian life, or its relics, might be found. ALH 84001 was not a surface rock. It came from underground.

Long before ALH 84001, it was clear that the possibility of life, even fossilized life, on Mars was the most exciting of the planet's mysteries. The fact that seeking out and interpreting fossils is a very hard thing to automate has always been one of the key arguments in any case for humans on Mars. If the exploration has to be done underground, the case becomes stronger still. There may be robotic ways to choose a site from which to explore the underground. Infrared scanners might be able to pick up the faint surface warmth that betokens an active hydrothermal system below. Radar used from orbit or from an aircraft should be able to reveal liquid water aquifers to drill into, if there are any, or subterranean ice deposits worth sampling. Aircraft with chemical sensors might be able to sniff out minute traces of gases produced by subterranean life; any methane or formaldehyde produced by underground anaerobes would last only hours in the Martian atmosphere, and so would tend to be found around vents that connected the surface to the life below.

*There's a wry little story by Stan Robinson, written in the form of a set of abstracts to articles in scientific journals, which suggests that even after people have been able to study the site on Mars from which the meteorite came they still won't be able to agree. "Selected abstracts from *The Journal of Areological Studies*," in *The Martians*.

Once you have found a place from which to start your underground explorations, though, whether it be by means of a drill rig or—a scenario to delight the imagination of Penny Boston—by actually exploring a cave system, you will have reached the limits of today's remote sensing technology. Serious drilling in unknown territory requires a drilling crew; in this, if in no other particulars at all, the movie *Armageddon* was right. And caving requires speleologists. If you're going to usher in a new age of Martian exploration with a serious attempt to understand what lies beneath the surface of Mars, you're almost certainly going to need to send people to do it.

Bob Zubrin's Frontier

> Thus in the beginning all the World was America.
>
> —John Locke,
> *Second Treatise on Government,* Sec. 49

Robert Zubrin is a short, intense and funny man with a somewhat wild air; he has a rapid-fire delivery, a frequent grin, and a rather fierce scowl. In 1990 he came up with the best engineering solution to the problems of getting humans to Mars that anyone had ever contrived. Since then, he's been hard at work trying to sell it—not on the basis of science, but on the basis of history.

Zubrin took a degree in math and science at the University of Rochester toward the end of the 1960s and became a teacher—he ended up as the on-the-road tutor to the Brooklyn Boys Chorus. A child of Sputnik and Apollo, he tried to fill his young charges with his love of science. He told them that no one did more for society, or was more worthy of respect, than scientists and engineers. If that was so, asked one of the kids one day, why was Zubrin just a teacher. Zubrin came up with an answer—he always has answers—but he took the question to heart. He began applying to graduate schools and the University of Washington's Department of Nuclear Engineering offered him a place. So, like many of his countrymen in years gone by, Zubrin redefined himself by heading west. In the summer of 1983 he packed his belongings into a decrepit Chevy Nova and drove across the country to study nuclear fusion. In 1988 he moved to Denver to work with Benton Clark at the aerospace company Martin Marietta.

A key reason for that move was that in 1987 Zubrin attended the third of the Case for Mars conferences in Boulder, Colorado, and got bitten by the Mars bug. Case for Mars III was a coming-out party: For the first time, the serving NASA administrator was there and so were a lot of people from aerospace contractors. The Paine report on "Pioneering the Space Frontier" had been published and, perversely, the *Challenger* accident had rekindled dreams of the space age as an age of adventure. There was a feeling that the world was coming round to the Mars Underground's way of thinking. And there was a clear understanding of the biggest problem facing manned Mars missions: the cost of getting into space in the first place. It was assumed—as it had been since von Braun's outline of a "Marsprojekt" in the 1950s—that the spacecraft on which people would first travel to Mars would be assembled bit by bit in orbit. That would mean a workforce in orbit and a cheap way of getting supplies up there. Unfortunately, neither was available. The journalist Gregg Easterbrook, who bravely confronted the conference with a talk entitled "The Case Against Mars," pointed out that at 1987 prices it would take something like $15 billion simply to lift a Mars mission's propellants into orbit around the Earth. That made the overall figure of $40 billion that the enthusiasts liked to throw around look suspiciously small. The space station program, announced with a budget of $8 billion in 1984, had already grown into a $32 billion project; no scientific case would convince Congress to start another project on a similar scale.

Between the 1987 conference and its 1990 successor, Easterbrook's analysis was pretty well borne out. In response to Bush the Elder's call in 1989 for America to go back to the moon and on to Mars, NASA had come up with a plan that called for vast new amounts of infrastructure, a greatly enlarged space station, new nuclear rocket systems, and God knows what else. NASA never put an official cost to this plan, but other people did and a figure of $450 billion started to be heard in Washington. You didn't have to be a Beltway insider to see that this was a problem; Bob Zubrin, a politically unconnected engineer sitting in Denver, was well aware that such a price tag made the proposal impossible. He could also see that the system it applied to was ornate and over-specified: It was a system designed by people more interested in the general possibilities of spaceflight than in the specifics of getting to Mars.

In 1990 Zubrin and his colleague David Baker were part of a team that Martin Marietta set up to look at alternatives to NASA's Mars plans. They needed something that was simple and quick—one of the biggest factors in a space program's costs is the time it takes. Their answer was "Mars Direct," a mission design that did away with all the complex and expensive on-orbit assembly in NASA's plans, and that didn't require any new nuclear propulsion technologies. Instead, it called for a large launch vehicle and for a way of making rocket fuel on Mars.

Mars missions like the ones NASA was talking about would be required to carry to Mars all the fuel they would need to come back: Fuel with which to return would, in fact, account for more of the mass the mission would have to send to Mars than anything else. The idea that things would be a lot easier if the fuel could be made on Mars was an obvious one, and it had been much discussed among the Mars Underground and beyond: Benton Clark, early inspiration of the Underground and Zubrin's boss at Martin Marietta, was particularly keen on the idea. But it had always been seen as something for later, something that would make getting back from Mars cheaper once a base had been established there. To send the first crew out with a system for making fuel for their return, rather than with actual fuel, seemed much too risky.

Zubrin found a way to reduce those risks by taking care of the journey's return leg before the outbound leg took off. In effect, he designed the mission backward. The Mars Direct plan called for a new class of rocket that would use derivatives of the space shuttle's engines to launch payloads as large as those of Apollo's Saturn Vs. Such rockets would be able to send about forty-five tons directly from the surface of the Earth to the surface of Mars, and each Mars Direct mission would require two of them. The first payload sent to Mars would consist of a small nuclear reactor, a chemical processing plant, a tractor, and a two-stage rocket capable of bringing a crew of four back from Mars to Earth. The rocket's tanks, though, would be almost empty. Instead of the hundred tons or so of fuel and oxidant that the spacecraft would require for its return to Earth, they would contain a small amount of hydrogen and nothing more.

When this payload landed on Mars, the tractor would pull the reactor away from the landing site, perhaps into a small crater. There it would start producing electricity for the chemical plant. The chemical plant would then use the hydrogen shipped up from the Earth to reduce carbon dioxide from the Martian atmosphere. That reaction (the same as the one that allows the bacteria in the Columbia River basalts to eke out a living) would produce methane, which is a pretty good rocket fuel, and water. The water would be broken down into oxygen and hydrogen using electricity from the reactor. The oxygen would be stored for eventual use in the rocket's engines as an oxidant for the methane; the hydrogen would be recycled to make more methane. Adding an extra set of chemical reactions in which hydrogen reduces carbon dioxide to carbon monoxide would allow yet more oxygen to be made. Within about six months of landing, six tons of hydrogen would have been turned into a hundred tons or so of methane and oxygen, and a fully fueled ship—an Earth Return Vehicle, or ERV—would stand ready and waiting on the Martian surface.

Two years later the second and third launches would take place. The second launch would carry a payload essentially identical to the first: reactor, chemical plant, tractor, and ERV. The third launch would be of the crew—four people—and their "hab": A spacecraft capable of sustaining them through the flight to Mars, landing them on the Martian surface, and sustaining them there for a couple more years. Though the hab would be launched a few weeks after the second ERV, it would follow a faster trajectory and get to Mars sooner. Once there, the crew would land it next to the fueled-up ERV waiting for them on the surface. Shortly afterward the second ERV would arrive from Earth and start fueling itself up for the benefit of another hab to be launched two years later. Having the crew launch later but travel faster was a little detail that provided a lot of extra safety. The crew wouldn't launch until there was a fueled-up ERV on Mars and another on the way. If the crew were forced to land far from the original ERV, there would still be time to direct the one in transit to come down close to them.

With an ERV and a hab launched at every opportunity, Mars Direct offers an ongoing Mars program for an average outlay of one vehicle launched a year and with no need for construction crews in orbit. Zubrin imagined that the hab, the ERV, and the big new booster could be kept cheap if they could be kept simple; he imagined the overall cost of developing the technology and launching the first few missions at about $30 billion. If each ERV was sent to a new site, a broad swathe of Mars could be explored over a decade or so: Zubrin imagined ground vehicles that would allow the crew to travel up to five hundred miles as a way of filling in the gaps between landing sites. Alternatively, a Mars base could be built up hab by hab at a site of particular interest: a borehole over a deep aquifer, say, or a young lake bed, or the entrance to a cave system.

By the early summer of 1990 Zubrin was presenting Mars Direct to engineers and managers at NASA. Carol Stoker heard him speak at Ames and was excited enough to give him a bully pulpit at that summer's Case for Mars IV, back in Boulder. The presentation was a hit. The Mars Underground had always been keen on the use of indigenous resources; Zubrin and Baker had found a way to make that part of the process from the beginning. That not only made their mission cheap. It made it compellingly Martian. They were not designing all-purpose spacecraft; they were designing the best possible way of getting to Mars. And this was because, unlike many at NASA, Zubrin was not primarily driven by a love of space travel for the hell of it. He saw space travel as a way of achieving his goal. And his goal was Mars.

Zubrin is as passionate about Mars as a place as anyone on Earth. The nature of that passion is neither sentimental nor scientific. It is ideological. Zubrin believes fervently in the frontier theory of history first enunciated by Frederick Jackson Turner in an 1893 talk called "The Significance of the Frontier in American History." Turner held that the uniquely impressive aspects of America's national character—a love of the new, a braveness in the face of the future, a willingness to adapt and grow—derived from the existence of a frontier to the west.

That frontier, Turner argued, was in the process of closing down for good and he wondered whether America could continue in its greatness without it. The wondering has gone on in some circles ever since—accompanied by suggestions that some other phenomenon might take the frontier's place. In the 1940s Vannevar Bush suggested in a report to President Truman that science itself could be an "endless frontier." In 1960, President Kennedy made "new frontiers" and their reinvigoration of society the central theme of his election campaign. The television producer Gene Roddenberry combined the two ideas into a "final frontier": *Star Trek*'s conflation of physical exploration with the search for knowledge was profoundly influential, making space travel the dominant metaphor for the creation of scientific and technological novelty. Space enthusiasts have been using frontier imagery relentlessly ever since. As the historian Patricia Nelson Limerick has pointed out, this use has been consistently self-deluding and ill-informed, but that hasn't put an end to it.* The computer age became yet another new frontier when William Gibson's idea of "cyberspace" gave it a sense of dimensionality; its pioneering freedoms of expression were upheld by an Electronic Frontier Foundation that had frequent recourse to the rhetoric of the Wild West.

I'm fairly sure Zubrin would laugh in scorn at the idea that a computer screen could be a frontier. Following Turner more literally than most, Zubrin believes that the frontier America needs is an idea that is also, first and foremost, a place. Such a frontier can be spiritually, technologically, and politically liberating only to the degree that it engages its inhabitants through physical constraints and privations. "There is no tabula rasa," Turner wrote in a passage that Zubrin was to cite. "The stubborn American environment is there with its imperious summons to accept its conditions." To Zubrin the imperious summons now necessary can only come from Mars. While others see space itself as the frontier, it's not placelike enough for Zubrin: Space is the way to get to frontiers, not a frontier itself. And of the places that can be reached, only Mars has the resources with which technologically adept humans could achieve self-sufficiency. It has water—not necessarily much, by planetary standards, but quite a lot by the standards of frontier settlements;

*In *Space Policy Alternatives,* ed. Radford Byerly.

the north polar cap would keep them going for quite some time. There's an atmosphere that can be "mined" for raw material; there's regolith from which you can extract minerals and make the bricks with which to build your homes and halls. There's sunlight enough for gardening in greenhouses. There is a vast array of practical difficulties inherent in actually making use of all this—but that's the point.

Zubrin's passionate commitment to a Martian frontier is inseparable from his belief in the increasing decadence of the Earth. "We see around us," he was to write in an essay called, in homage to Turner, "The Significance of the Martian Frontier,"

> an ever more apparent loss of vigor of our society: increasing fixity of the power structure and bureaucratization of all levels of life; impotence of political institutions to carry off great projects; the proliferation of regulations affecting all aspects of public, private, and commercial life; the spread of irrationalism; the banalization of popular culture; the loss of willingness by individuals to take risks, to fend for themselves or think for themselves; economic stagnation and decline; the deceleration of the rate of technological innovation . . . Everywhere you look, the writing is on the wall.

From the summer of 1990 on, Zubrin's anger at the status quo and passion for the future made his Mars Direct pitches much more than just talks about engineering: They were sermons with viewgraphs. Zubrin's case for Mars was a societal and moral imperative, the need to revive in the twenty-first century a history that had ground to a halt at the end of the nineteenth. Going to Mars was a crucial opportunity. If that opportunity were seized, it would become the preeminent achievement of our age, the thing for which we would come to be remembered and honored.

Zubrin mapped out much more than a mission strategy in his talks: He mapped out a first draft of Martian history. He argued that Mars Direct could lead to the founding of a permanent base, then of a colony, then of a new branch of civilization. As the talks went on, graphs showing rocket engine performance gave way to graphs showing balances of trade and population growth as Zubrin warmed to the theme of bringing Mars to life. In later centuries there would be farther frontiers, and humanity would spread its seed across the galaxy. And it would all be due to the decision to go to Mars that could be made right here, right now. Echoing Pericles's oration for the dead of the Peloponnesian War, he told the audiences that future ages would wonder at them, even as the present age wonders now—and that the wonderers of that future would be spread across the breadth of the Milky Way, forever inspired by the select few who first expanded the human world to a second planet. The mixture of nuts-and-bolts engineering with historical sweep and more than a touch of wild-eyed zealotry was an intoxicating one.

For all Zubrin's fervor, though, Mars Direct was not able to revive the president's call for a mission to Mars as a national goal. Even if NASA had embraced Mars Direct immediately, Zubrin's and Baker's plan would have been unlikely to make a difference at that late stage. And NASA did not embrace it. In part this was because the space agency was much more concerned with the space station in its hands, which was already feeding prodigious amounts of money into the coffers of its client-contractors,* than it was with Mars missions in the bush (this was part of the reason that the agency's overall response to President Bush's call was so lackluster). In part it was because Zubrin made no bones about how stupid he found the conventional mission models that relied on huge ships—"death stars," he called them—built in orbit. Attacking a bureaucracy's vested interests and its technical proficiency simultaneously is a high-risk strategy.

Objections thrown up against Mars Direct—that it requires the astronauts to be subjected to too large a dose of radiation or too long a period of zero gravity; that a crew of four is too small; that surely it makes more sense to assemble things at space stations, or mine materials from the moon—were batted away by Zubrin. Gravity would be provided by spinning the spacecraft at the end of a tether; radiation risks were overblown; explorers had been far more isolated in the past. Not all these arguments convinced the critics,

*Perhaps not coincidentally, Zubrin's reasonably indulgent employers at Martin Marietta had less of a stake in the space station project than any of the other major aerospace companies.

but they were strong enough to make the objections feel like details to be dealt with rather than showstoppers. Eventually parts of NASA overcame their not-invented-here reflex and embraced a modified form of Mars Direct as the agency's Mars "reference mission"—the mission design against which other proposals should be measured. But that was not until 1993, long after the idea of humans to Mars had vanished from the realm of political discussion and the question of how to mount such a mission had become safely academic.

Zubrin, though, did not give up. He became chair of the National Space Society, an activist organization. With the help of Richard Wagner, a journalist who edited the National Space Society's magazine, he wrote up his ideas as a book, called *The Case for Mars.* And when hundreds of readers wrote to him eager to become Periclean heroes to future generations, Zubrin set up a new organization of his own, the Mars Society.

The founding convention of the Mars Society took place in Boulder in 1998. In some ways it was a successor to the series of Case for Mars conferences, which had fallen on hard times. They had started off small, tucked into little rooms in the basement of the university's arena; as interest had grown they had swelled until they filled the Memorial Center, the university's main public forum. But by 1996, with George Bush's pretensions as exploded as *Mars Observer* and no real progress to report since *Viking*, they were back to the little rooms under the arena.

The Mars Society convention retook the Memorial Center in high style, bringing in perhaps seven hundred people, mostly from America but some from farther off, most of whom had paid their own way to be part of the dream. There were Air Force cadets and a Lutheran priest (James Heiser, now semi-official pastor to Mars), Internet nerds and nuclear engineers, social scientists and artists, and science fiction writers and philosophers. The core of the Mars Underground was there, as were a number of scientists and NASA types, but many of the attendees were not particularly technical; they were just convinced that Mars was the next big thing and they wanted to be part of it. Over a beautiful weekend in a beautiful city they shut themselves up in the meeting rooms for nine or twelve hours at a time to talk about all things Martian, from schemes for financing missions to the names of the new months that would be needed for a 669.6-day Martian calendar, from the design of tractors for regolith-dozing to the ethics of changing the Martian environment.

Wandering around that meeting, it was impossible not to be struck by the sincerity of the people attending; but it was also impossible to see quite how they might go about achieving their goal. I first heard Zubrin's Mars Direct pitch at the 1990 meeting of the International Astronautical Federation in Dresden.* At the same meeting Bruce Murray gave an impressive speech asking whether the end of the Cold War was also the end of space exploration. The argument was a strong one. Superpower rivalry had, after all, been the main driving force in space exploration. The technology for launching things into space came from the technology for throwing warheads around the Earth. The Apollo missions had been a race for respect in which the United States had come from behind to win. When there was no overriding national aim, space programs stagnated into job-maintenance programs like the shuttle.

At the end of the 1980s there was some talk of inverting the historical relationship between space and the Cold War by cooperating instead of competing: Gorbachev's glasnost-giddy Soviet Union, the argument went, could join with America to mount a Mars mission. Murray's friend and sparring partner Carl Sagan championed the idea under the slogan "To Mars—Together." The Soviets brought it up (briefly) at summits. But the American government was never convinced, and as the Soviet Union collapsed so did Sagan's dream.†

*It was a meeting at which NASA's talk of going to Mars was somewhat undercut by the fact that budget wrangling meant it could only just get half a dozen speakers across the Atlantic.

†In the 1990s similar thinking was one of the things that led to Russia's inclusion in the development of America's space station (in which Europe, Japan, and Canada were already junior partners). It was a piece of symbolism intended to give the project an added purpose and to provide gainful employment for Russian technologists who America believed might otherwise sell their know-how to insalubrious customers in the developing world. The process has brought some benefits, but the project has been rife with delay and cost overruns. It is not clear that this would be a good way to go to Mars.

At present there is no pressing political need for a major new space program, and history suggests that, without such a need, manned space programs languish.

Zubrin's ideas about reopening the frontier are compelling to many members of the Mars Society, but there are two problems inherent in trying to make them sound like an urgent national need. The first is that the frontier thesis, while undeniably widely held, is also widely questioned. Historians have spent a century pointing out the many weaknesses in Turner's arguments: that he ignores the native Americans; that he overemphasizes the role of individual homesteaders and plays down the importance of industry and capital; that he unreasonably conflates the settlement of the Great Plains with the opening of the mountains and deserts; that he plays down the role of federal power, of Chinese indentured labor, of worldwide changes in industrial manufacturing, and of East Coast political intellectuals. Admittedly, this revisionism may have had relatively little impact outside academia—as Patricia Nelson Limerick points out wryly, Turner's ideas have the status of a certified classic and, like cars that achieve the same status, they are allowed to flout the emissions standards applied to other vehicles—but it still counts for something.

More important to the political potential of the frontier argument, though, is its overall tenor. The necessary corollary of buying into Zubrin's argument for a frontier is a belief that there is something wrong with the frontier-free status quo. To tell people that a Martian frontier is needed is to tell those who think they are doing okay that they are mistaken and to play up to the resentments of those who think the world is against them. There was a lot of rather sweet idealism at that first Mars Society convention, but there was also a fair amount of resentment. Some of it was the resentment of a child denied: These were people who, growing up in the 1960s, had been told there would be a space age and they wanted that dream to be delivered. Some of it was more general and more bitter: the perverse resentment of people who believe that the society that continues to accord them a privileged position is engaged in a plot to stifle them and hold them back.

Even for a sympathetic observer like me,* there was something a little distasteful about a group of the world's most fortunate people—the vast majority of them middle-class white male Americans, safeguarded from risk by background and education and insurance and any number of other, subtler barriers—complaining that a planet where a billion live on less than a dollar a day and millions die from easily preventable diseases is too mollycoddling and risk-averse and constraining. When you hear someone saying (of future Mars colonists), "It's a frontier. People are supposed to die. That's the point," it's hard to resist suggesting he just go and play in the traffic—or work in southern Sudan—until his appetite for risk wears off. People who argue like this may, for the most part, actually believe that having their fantasy made real will improve the world for the poor and suffering that current prosperity leaves behind (the "people are supposed to die" man turned out, when I talked to him, to be sweet and smart). But they can hardly make a case that it's the best way to do so—or the quickest.

These resentments add an ironic twist to the Mars Society's attitudes toward government. Their mindset is largely that of people who want government to get out of their way. But in practical terms the government is not *in* the way: It *is* the way. No one will get to Mars without some government or other footing most of the bill.

Many people at that first Mars Society convention, and since, have tried to find a way around this fact. They've talked of sponsorship deals with media companies, of private philanthropy, of bond deals financed on the basis of future earnings from Martian resources. For some, private enterprise is not just an alternative means to the desired end; it is an end in itself. The Apollo program stands out as evidence that there are some things only a government can do and as a suggestion that there is no limit to the power of such collective action: Ambitious programs are still pitched on the basis that "if we can put a man on the moon . . ." nothing is impossible. Leaving aside the fact that many things have proved impossible for the

*Rather priggishly, perhaps, I waived the opportunity to become a founding member of the Mars Society on the basis of a need for objectivity in reporting on it. But I don't imagine you will be shocked to hear that if they get their wish I'll be delighted.

nation that put men on the moon (including continuing to put men on the moon, or putting women there), there is a certain libertarian cast of mind that finds this sort of faith in the power of government dangerous. Entirely privatized space projects that overtook Apollo's achievements would improve the world simply by humbling government's last redoubt.

Whatever you think of this line of argument, it has one drawback as far as most Mars Society members are concerned: It's not going to put people onto Mars any time soon. Sponsors are not going to provide billions of dollars up front for a venture that risks failing spectacularly. And there's nothing you can realistically imagine finding on Mars that would equal the expense of getting there in terms of cash. Indeed, this is perhaps the greatest shortcoming of Zubrin's frontier thesis. The various parts of America that at one time or another constituted its frontier were "opened" by economic activity, sometimes state-sponsored, sometimes not. People went to the frontier to mine, to farm, to build railroads. Though there may be minerals on Mars that might, under certain circumstances, be sold on Earth for more than the cost of shipping them back, that does not mean their value would be worth the capital costs of setting up the infrastructure needed for their extraction and transportation. The same argument applies with far greater force to agricultural products.* As to transport, the railroads did what they did for the West because it happened to sit between the two prosperous coasts. Mars will always be the end of the line.

This is not to say that people will not eventually visit Mars, or indeed colonize it. In the foreseeable future, though, it will not be on the basis of private profit. Nor, I suspect, will it be on the basis of reviving a civilization they fear is in decline. In fact, I suspect it will be the reverse. The American government could easily afford a Mars mission; in all likelihood it would cost little more than the space station. The problem lies in the symbolism—in spending all that money on a nonproblem when there are plenty of real problems

*Paul McAuley, in his novel *The Secret of Life,* imagines that Martian life-forms might be of immense financial value to the biotechnology industry, which is an intriguing idea but not a compelling one.

to be dealt with. Zubrin's argument that money spent on Mars would indirectly help to solve many other problems may contain some truth. But it is one that interest groups seeking money for more obvious and well-entrenched purposes can easily dismiss.

When attacking the idea that going back to the moon is a necessary prerequisite of going on to Mars, Zubrin is given to quoting former NASA administrator Thomas Paine's quotation of Napoleon Bonaparte's advice to a general whose strategy was too convoluted, "'If you want to take Vienna, take Vienna.' Well, if you want to go to Mars, go to Mars." The same sentiment, though, can be turned back against Zubrin's main argument that Mars exploration is a way of bringing about desirable societal change. If you want to revive a world you find unsatisfactory, work directly to revive that world. Work to make it fairer, more democratic, more confident in its abilities to solve its problems.

These are things worth doing in and of themselves. But they are also, I think, ways of building the sort of society that will be most likely to undertake the human exploration of Mars. Mars missions are more likely to be launched by a confident society already convinced of its capabilities than by a society trying to fight off its perceived decadence. They will be a product of that society's spirit, and quite possibly a powerful symbol of it, rather than its cause.

If the world grows richer, fairer, more technologically accomplished, there will come a day when the original case for Mars—that to explore it might be a great scientific bonanza and would certainly be a great adventure—will be enough in and of itself. The question will shift from "Why go to Mars?" to "Why not?" And when that happens, we shall be on our way.

Mapping Martians

> Oh man, look at those cavemen go.
> It's the freakiest show.
>
> —David Bowie, "Life on Mars"

Since the founding convention in Boulder in 1998, the Mars Society has been busy. Its chapters have lobbied for the funding of robotic Mars missions; there have been annual conventions, spirited arguments over the right way to proceed and some serious fund-raising. There has been work on the development of spacesuits and rovers; the Australian chapter has secured space for a dedicated radio pay load on a small experimental satellite. Zubrin's company, Pioneer Astronautics, has worked with NASA on the design of Mars balloons—aerospace as agitprop, since Zubrin believes, as do many others, that pictures of Mars from the air will make the planet more interesting to earthlings and stoke up the fires of exploration. There have been fallings-out and resignations from the board. Most strikingly, there has been the Flashline Mars Arctic Research Station, a dummy version of one of the Mars Direct habs. It sits on a lifeless-looking ridge on Devon Island, high above the Arctic Circle in the semiautonomous Canadian territory of Nunavut. As I write, it has survived its first winter and its first fully crewed summer.

Devon Island is a partially glaciated wilderness with no permanent settlements on it, not as profoundly arid as the dry valleys of Antarctica but still dry enough to be classed as a desert. Near its northern shore sits Haughton crater, about twelve miles across and twenty-three million years old. When it was first spotted during aerial surveys in the 1950s it was assumed to be a salt dome; in the

1970s Canadian geologists realized it was actually an impact crater. If such an impact were to occur today it would be a global catastrophe: The dust and debris thrown into the stratosphere might chill the climate enough to ruin crops for years and precipitate global starvation.

Low precipitation and scarce vegetation mean that Haughton is in some ways quite fresh: Although there has obviously been a lot of erosion over twenty million years, many of the rocks that remain are much as the impact left them. There is no other crater of its size remotely as well preserved. In the 1980s it was studied quite intensively by a joint German–Canadian–American project. Then, in the 1990s, a young researcher named Pascal Lee realized that, as a large, well-preserved impact crater in a cold desert, Haughton represented one of the most Martian locations on the Earth. With impressive determination, Lee turned this insight into the NASA-funded Haughton-Mars Project (HMP), which is run out of the Ames Research Center. The Mars Society's activities in Haughton now run in parallel with the HMP's main scientific program.

Part of that program is studying Haughton. The other part is studying how you go about studying somewhere like Haughton, and looking at how such fieldwork might be carried out on Mars. There are geologists and biologists there, but there are also technologists working on robotics and new computer interfaces, and technology development shades into anthropology as the computer people document the culture in which the research gets done. Bill Clancey, an expert in "human-centered computing" at Ames who has been studying the people studying Haughton since 1998, puts it like this, "You describe and think about the particulars that most people take for granted—how the camp is laid out, how the space is used, how the daily routine is planned, how people are showing each other and talking about rock and geological formations, how results are shared over dinner conversation and so on." In other words, you try to capture the researchers' world in all its complexity—because it is a world like this, not just a bunch of people and supplies, that a Mars mission must try to transplant to another planet.

Over the years, Clancey has taken on the role of an anthropologist watching the development of Haughton's rituals and routines. He's recorded the way that different tents get used and the role that e-mail plays in how people organize their lives. He's noted the way that the garaging area for the little All Terrain Vehicles (ATV) that are used to get around the crater has become the camp's main meeting place. (Part of the appeal lies in the ATVs' seats, which are the most comfortable lounging sites around when stationary—when moving over the broken ground at speed they are as uncomfortable as you would expect.) He's learned that most people don't know who does the washing up in the morning. He's learned the extent to which scientists in the field plan their work around the paper they are already drafting in their heads, and the degree to which everything they do is a response to the tools that they start with and the ways those tools can be adapted.

He's watched as more and more of the crater's features have been named. Some naming is serious: The ridge that the Mars Society's Flashline hab stands on is called Haynes Ridge in memory of Robert Haynes, a Canadian biologist with a strong interest in Mars who died in 1998. The plain beyond the landing strip is called Von Braun Planitia in honor of the author of the Marsprojekt. Some naming is commercial—the Flashline hab is so called because Flashline.com sponsored it—and some is dutiful: The lakes are named after universities and colleges—Lake Stanford, Lake Cornell—though there is also a Lake Astrobiology Funding, so called because it dries up halfway through the year. Water for the base camp comes from Lowell Canal, and the dried lake bed near the center of the crater that offers richer soils and thus more vegetation is Lowell Oasis. As things get smaller, their names get sillier: *Pathfinder*'s cartoon-based naming tradition starts to appear through the cracks in the IAU's more formal protocol. There's Devo Rock, named for the early 1980s band of the same name (their guitarist, Frank Schubert, built the Flashline hab); there's Dr. Evil's Lair. Indeed, Austin Powers is a pervasive presence. The little rubber dinghy used for sampling the lakes is the Research Vessel *Mini-Me*.

In the summer of 2000 the addition of the Mars Society's Flashline Station, a two-story cylinder that can house six people, raised the Mars simulations at Haughton to a new level; but it started off with near disaster. The hab's parts were dropped in installments by a Marine Corps transport plane. Installments one to four came down fine; installment five's parachute failed to open. The hab's floors and a crane and trailer needed for construction were ruined, and most of the construction crew quit. As soon as he heard about the disaster, Zubrin flew from Denver to Resolute, the nearest town to Haughton. He and Schubert put together a jury-rigging package of wood and tools, commandeered an airport luggage cart, and flew out to the crater. With determination, imagination, ideal weather conditions, and help from local Inuits they assembled the hab's cylindrical walls, cinching them together with a makeshift belt when there turned out to be gaps between them; put on a roof; and set about making plywood floors. Triumph was snatched from adversity with a resourcefulness that left Zubrin aglow. This was the true spirit of the frontier—what robots could have done such a thing?

For a few brief days, Mars Society members occupied their outpost under the command of Carol Stoker and imagined themselves on Mars, to the extent that such a feat was possible given the noise of the continuing carpentry. They discussed their plans with "mission control." They treated their trips outside the hab as Extra-Vehicular Activities (EVAs—a crucial bit of NASA speak) with associated radio protocols and safety rules. They filed reports on the Internet; they gazed out of the window. This last was oddly moving; from the window, no sign of humanity was visible in the ruined landscape. "I think the Devon Island location looks the most 'off world' of any place I have been," says Stoker, "and I have seen more sights than most people. Looking out the porthole frames the view in a unique way. The scene is very beautiful, but is strange enough to make it easy to believe that one is not on Earth any more."

If it sounds like a game of make-believe, well, in a sense so it was. So is the Mars Society's competition to design a rover simulator that could also be used for such studies, and so are its plans for a communications relay on an Australian satellite that will be used, if all goes well, by Flashline's future inhabitants. They are make-believe and they are fun (Clancey has noted that a commitment to having fun is one of the things that Haughton's culture prizes most highly). But, at the same time, they are serious. They are serious for the Mars Society because they are one of its main conduits to the media. Haughton in general and the Flashline hab in particular were covered quite extensively in the summer of 2000. There may also, in time, be a revenue stream—Zubrin has plans for hab franchises in all sorts of deserts, starting in the American Southwest, that people might pay to visit. The games are also serious for the participants. Seeing the desolation of Haughton through a spacecraft window was a powerful experience. And they are serious for people like Lee and Clancey. If Haughton is to teach people about how to explore Mars, then ways have to be found of exploring it within Martian constraints. The Flashline hab is one of those ways. In time such buildings will be to would-be Martian explorers what the vast swimming pools at Johnson Space Center are to would-be astronauts: the best available Earth-based simulators for unearthly experiences.

Such simulations are intended to teach people what working on Mars will be like. Already, one answer is clear: Given current technologies, it will be frustrating. One of the key arguments for sending people to Mars in the first place is that human fieldworkers will be far more capable than robots ever can be. It was a point made very neatly in field trials that took place in February 1999, during which a team at NASA Ames tele-operated a rover carrying various instruments slated for Mars Surveyor missions. The rover was placed in the Mojave Desert, but treated as though it were on Mars. The rover team was reasonably accurate in many of its assessments, but not nearly as good as geologists able to survey the area on foot. If the geologists had been in spacesuits they would still have outperformed the rovers—but it would have taken a lot more time and effort. Spacesuits are heavy and cumbersome; they take time to get in and out of, and they limit the time you can spend on a given task. They cut down your peripheral vision; their faceplates may well get covered by Martian dust. They remove your sense of touch almost completely; they make working in tight corners or on steep slopes difficult if not impossible. They add to your weight and they malfunction. They cut you off from the world in which, as a receptive scientist, you are meant to be immersed.

Part of the point of simulating Mars on Earth is to understand these limitations. Another part is to work out ways around them. One idea is to minimize the amount of work that needs to be done in spacesuits in the first place. At the second Mars Society convention in 1999, the film director James Cameron held a capacity audience enthralled with production designs for his forthcoming Mars production (still forthcoming as of this writing). Cameron is, as his films suggest, a world-class gadget freak, endowed with a fertile engineering imagination and a prodigious capacity for absorbing and caring about technical details. If he was going to design a Mars mission, he was going to do it right. One of his more striking insights was that the traditional portrayal of Mars rovers as vacuum-proof Winnebagoes designed to move people around might be misleading. Given the problems and risks of spacesuit work and the time taken up by preparing for even a simple EVA, a lot of fieldwork might actually be done from within the rovers by means of remote manipulator arms. The rovers would be like wheeled versions of the deep-sea submersibles he had used to film *Titanic*, mobile bases for the control of remotely piloted vehicles, with samples passed in to the researchers through small airlocks.

Another way around the problems of Martian fieldwork is the design of better tools. If a geologist cannot use a hand lens in a spacesuit, then the suit's visor will have to magnify things. If she cannot wear a knapsack and a spacesuit, maybe some sort of robot carrying cart will have to follow her around. If she cannot scribble notes or tap on a keyboard handily (protecting a keyboard from corrosive Martian dust will be quite a task), various forms of voice recognition software will be necessary. If she cannot sketch a map with her hands in gauntlets, she will need a digital camera that somehow lets her annotate its images.

None of these tools, though, will be freestanding. Haughton crater is already the site of a number of computer networks, and each season adds new connectivity. The Internet is not an alternative to scientific exploration; it is a new facilitator. Wireless systems keep the Haughtonites in touch with each other and with various roving and flying robots. Samples are logged, pictured, and discussed with colleagues continents away over the Web; personal e-mail provides insulation against the loneliness. The networks are not only a way out; they're also a way in. The HMP Web site offers beautiful virtual reality panoramas for Internet visitors. The Flashline hab may not have had a proper floor in its first summer, but it had three cameras trained on its occupants—one for the Discovery channel, which had a sponsorship deal, one for the Mars Society Web site, and one for Clancey's anthropological studies*—not to mention a number of modems all of its own.

And this is nothing to what things will be like on Mars. Explorers on Mars will have high-capacity local-area networks linking their habitats, rovers, individual spacesuits, and even some of their tools. Spacesuits will ceaselessly update the network on their position and on what their occupants can see; remote cameras—perhaps on tethered balloons—will keep an eye on all movement. Data-mining programs will work on reconciling human observations and annotations with remote-sensing data from other sources to suggest fruitful targets for future excursions. To maximize the efficiency of the human explorers—who, on early missions at least, will always be too few and too pressed—everything that can be recorded will be recorded and everything that can be automated will be automated. Data will constantly be shared over the Web with mission control and earthbound scientists. If something terrible happens, constantly backed-up earthly databases will act as a virtual black box.

No hill or ridge or crater rim will ever be crested without foreknowledge of what lies beyond. The explorers standing on the desolate surface of Mars, as physically isolated as any human has ever been, will at the same time be immersed in a world of human data. They will know where they are and where their companions are. A word whispered into a microphone will bring them images of where they have been, or of where they are going. If one of them so much as looks at an intriguing outcrop the discrete heads-up display projected onto her suit's visor will tell her if it has already been studied and, if so, what was found. They will be cocooned in data as snugly as they are wrapped in their suits. Their planet may be alien, but their world will be connected.

*Clancey himself was inside the hab.

Imagine the first landing.* Zubrin's approach is now mainstream enough for us to be sure that, wherever the site may be, it will already have been prepared in some way. There will be physical infrastructure—a fueled-up rocket ship, in all likelihood, along with a fuel factory and a power supply, perhaps a rover and a greenhouse. Radiolocation beacons will be ready to guide the ship to within inches as it sets itself down. If cameras on tethered balloons have not already produced images of the surface's every pebble, cameras in orbit will have done nearly as well. Our high-resolution models will tell us the thermal properties of the soil and the mineralogical composition of the rocks.

Up in orbit they will check their real-time data on the upper atmosphere, make any last tiny corrections to the flight plan, and fire the engines. The cameras within the landing craft will go offline; the crew will have negotiated its allocations of privacy carefully and this will be a time it will not want to be watched. The cameras on the landing craft's exterior will see the surface grow close and closer. Then the fires of atmospheric entry will blind them. High data-rate communications with the spacecraft—a constant part of our mediated earthly lives since its launch seven months ago—will be briefly severed.

Cut off from our explorers, as a world we will gaze up through cameras on the surface, straining to see the first signs of a parachute. Just when we think we never will, it will appear. Descent cameras on the landing craft's legs will show us the already familiar ground rushing closer; cameras on the surface will show us the spacecraft falling through the washed-out sky until it is lost in the dust kicked up by its engines. And then they will be there. We will hear a voice from Mars, and it will self-consciously echo the voice we once heard from the moon. It will tell us that what was once an uninhabited site is now a base and that a spacecraft whose name will long have been a household word has landed. The local network and the satellites

*And, if you don't mind, imagine it on a Tuesday. In Sanskrit Mars is Mangala, and the second day of the week is Mangalavar. The convention has carried through to Latin languages—Mardi—and to English—Tiu is another war god. If it's Mars, it must be Tuesday.

overhead will determine the explorers' position precisely. When we refresh our screens, their landing craft will have been incorporated into the terrain models.

Television anchors—if there are still anchors—will strive for something significant to say. Stan Robinson, Bob Zubrin, Matt Golombek, Chris McKay, and all the other specialists brought in to add to the commentary will be looked to for response; some will be lost for words. (How old will they be? I don't know.) Some of them may cry. Some of us watching will certainly cry. But it's only fair to say that the majority will be less strongly affected. To most it will be an entertainment.

There will be a pause—a pause long planned and scheduled. (On *Apollo 11*, the crew was ordered to take a nap after landing, though they didn't.) Then the first EVA will be prepared. We'll probably be able to watch some of the suiting-up; we'll see a face with which we have become familiar vanish under a helmet, its visor strongly tinted against the hard ultraviolet rays outside. We'll change our point of view and see through her helmet's camera.

The ship-side door of the airlock will close and she and her companion (EVAs will be in pairs; everything will be by the book) will be alone with the world. The air will be pumped out and the microphones on the suits will hear nothing as the pressure falls to a near vacuum. Then the outer door will open. There will be a noise of some sort as pressure equalizes; a little oxygen will escape out into the Martian air. She will walk forward, turn around, and start to climb down the cold rungs of the ladder. Her helmet camera will show us the side of the landing craft, already dirty with the dust kicked up on landing. Perhaps the microphone will pick up some faint noise as her feet hit the rungs; perhaps the thin wind will sing as it whips through the spacecraft's legs.

From the cameras outside we will watch her lower herself down, slowly, in the unfamiliar gravity. She will seem to be drawing the process out. Impatiently, we, or the networks, will flick between the two cameras with the best points of view, triangulating our impatience.

At last, one boot will touch the surface, its precise location immediately registered in the models. The fact that we will all know that this actually happened fifteen minutes ago will not matter at all. "When" may shift, today it is "where" that matters. Never has a location mattered to us more.

Her lessened weight—it's a small planet, after all—will shift from the foot on the earthly ladder to the foot in unearthly soil. Surrounded by absent observers she will step onto Mars.

And into our maps.

Part 5 – Change

But Manue Nanti still sat on the ground, his head sunk low, desperately trying to gasp a little of the wind he had made, the wind out of the ground, the wind of the future. [The supervisor] said nothing for a moment as he watched Manue's desperate Gethsemane.

"Some sow, others reap," he said.

"Why?" the Peruvian choked.

The supervisor shrugged. "What's the difference? But if you can't be both, which would you rather be?"

—Walter M. Miller, Jr., "Crucifixus Etiam"

Symbols of the Future

Love, Fly with me to Utopia:
three majestic snow-cowled volcanos
poking up through the sockeye dust.
Like Sherpas astraddle our mechanical
goats, we'll guide parties
all across the chapped terrain,
early seacliffs and ocher pastures,
tending our rock-leeches that suck
mineral and water till, gorged,
they thud like geckos to the ground.

—Diane Ackerman, "Mars"

Charles Cockell is a man of two poles. In the northern summer he heads up to the Arctic to work on the biology of Haughton crater; in northern winter he heads down to Antarctica to look at other sorts of extreme biology, including sensitivity to UV radiation. He'd like to go further. He's been fascinated by Mars since the age of seven. If a mission were heading off to Mars today, Cockell would be a good choice for the mission's biologist; he's easy to get along with, bright and enthusiastic, but with reserves of restraint and experienced in the ways of exploration. However, no mission is heading off, so the Martian adventures he would love will remain imaginary—or simulated. In the summer of 2001 he was in the crew at Flashline.

On the wall of his office at the British Antarctic Survey, on the outskirts of Cambridge, sits one of the most recent triumphs of terrestrial cartography. It is a strangely unfamiliar map of Antarctica: a map of Antarctica without the ice, compiled through radar measurements of the bedrock. In a pallet of greens, yellows, and blues similar to those of the MOLA maps of Mars, though more muted, it reveals secrets that have been hidden for as much as thirty million years. It shows the subglacial highlands and lowlands, the blanketed mountains, the basins that now contain lakes trapped beneath the ice, the broad swathes of rock that sit well below sea level, the vast channels cut by ice streams that so intrigue Baerbel Lucchitta. There are evocative names—the Gamburtsev Mountains, the Aurora Basin—but nothing more. No boundaries, no territories, no true places.

Cockell, who is boyish in a very British way, finds the map's mixture of fidelity and inaccessibility fascinating. He imagines its secret mountains and plains as the settings for stirring stories, "a sort of mixture between *Lord of the Rings* and *Swallows and Amazons.*" Maps of Mars demand stories in much the same way; the difference, odd as it may seem, is that stories based on the maps of Mars are far more likely to come true. The East Antarctic ice sheet, which covers the largest part of the hidden continent, has been in place for many millions of years. It will remain there for many millions more, shutting off access to the lands below other than through boreholes. Mars may be farther off, but millions of miles of empty space are less of a barrier than a few miles of solid ice. Mars is closer to us, in this way, than parts of our own planet are, and far more likely to become part of the human world. What are now stories about Mars could, conceivably, become histories. The Mars of the future may be different. The maps of Mars may have to change.

In one respect the maps of Mars are already changing, as more and more data from more and more sources are added, and this will continue for some time. Spectrographs, cameras, and magnetometers will map more and more of the planet at better resolution. Orbiting radars will begin to map the subsurface, as may seismometers on the ground; Robert Haberle's little weather stations may chart the weather. But though these processes will make our representations of Mars richer in data, they will not necessarily become richer in symbols. Maps of the Earth are covered with symbols to indicate where cities and borders and roads and airports and any number of other things are. On almost every map there is a key on one side explaining what these symbols mean. The only things worth their own symbols on general-purpose maps of Mars today are the landing sites, the only things on the surface other than its gross characteristics that everyone can agree on. The 1:15,000,000 USGS maps on my wall pick out the *Viking 1* and *Viking 2* landing sites with delicate reticules. (The less precisely known locations of the failed Soviet Mars landers of the 1970s are left off.)

In the years to come the landing-site symbols will proliferate, as more Martian locations become places. In early 2004 two little crosses will be added for the landing sites of two NASA rovers. By late 2001 four sites were short-listed, one in the hematite region of Terra Meridiani, one in Gusev crater, one in Athabasca Vallis, a channel in the very young part of Elysium, and one on the floor of Valles Marineris. Another cross, if all goes well, will mark the landing site of *Beagle 2,* a British-led European project similar to *Pathfinder* but with no rover and more instrumentation, which should land in the middle of the Isidis basin on 26 December 2003. In 2007 the four seismometer-bearing "Netlanders" of a French-led European mission are due to spread themselves around the planet.* In the same year the first of a new generation of smaller and cheaper (in principle) "Scout" missions should arrive. The year 2009 is expected to see a new, more capable American rover (it might get there earlier), and the same year or 2011 might see more Scouts and maybe even more Netlanders. The year 2011 might also see—at last—the first leg of the deeply deferred American–French sample return mission.

Not all the missions will be easily represented by landing sites; some may be airborne. Though the Martian atmosphere is thin, engineers have designed aircraft and balloons that should be able to make use of it, providing a new way to do science (for radar and magnetometers, performance from an aircraft could be far better

*Since most earthquakes are side effects of plate tectonics, one might expect there not to be any Marsquakes for seismometers to listen to; but any solid body will oscillate, if disturbed, and it appears that the Martian atmosphere, thin as it is, is capable of exciting these oscillations in the planet it encircles. The whole planet rings like an impossibly faint basso-profundo wind chime. On Earth such faint seismicity would be hard to measure, but on Mars things are a little easier. Earthly seismometers are limited in their sensitivity by the constant pounding of ocean waves on the Earth's shores. Even those stationed in the centers of continents are constantly aware of the sounds of the seas. Sea-free Mars is silent.

than from orbit) and a new way to turn bits of the surface into something like places. An aircraft's cameras would not just look down at the surface of Mars. They would look across the surface; they would look out to the horizon; they would insert changing perspectives into the view in a way that orbital images never could. And they would capture movement. They would see new features rise over the horizon and slip beneath and behind them. They might see the aircraft's own shadow sliding along the rocks below. They would catch the dip of the wing toward the ground as the aircraft carrying them slowly banked on to a new heading. There could be little better way to turn Mars from a planet being studied from afar to a world being experienced from within—if only briefly, if only vicariously. Not a little sandbox world like that of a rover, but a world of scope and scale and speed, a world like the one that can still impress the most jaded frequent flyer when it catches him off guard through the window of an airliner.

Mars might receive other visitors too. Russia might launch an independent Mars mission, if its planetary program is not fully integrated into the wider European program. The Japanese launched a Mars orbiter in 1998 that, due to a malfunction, is taking rather a long time to reach its destination (it should make it in 2004). A lander might be a logical follow-on. The Canadians have announced Martian ambitions. The Chinese might conceivably try their hand; so might the Indians. All in all, the coming decade could see at least a dozen missions to the Martian surface. Even assuming—as we probably have to—that a third of them will fail, that still means a handful of new landing-site symbols scattered over the maps. When networks of little weather stations and seismometers start to get dispersed over the surface by the dozen, their sites might even stop being marked on whatever passes for a general-purpose map in the GIS age. Getting to the stage where an earthly presence on Mars can be passed over will be an achievement in itself.

Around the same time more detailed reconnaissance might lead to a new sort of symbol being added: places notable not for a landing but for a landscape. Wind erosion carves some fairly remarkable features on Earth and it can be expected to do much the same on Mars. Indeed, Martian wind forms might be much more impressive in some places than the Earth's, given the low gravity and the lack of rain, chemical weathering, and earthquakes. Pamela Lee, an artist who worked with Bill Hartmann and Ron Miller on *Out of the Cradle*, a sequel to *The Grand Tour* that dealt with human exploration, has imagined a Martian version of Grand Arches National Monument, with sandstone spans like the ribs of a fallen titan. If such wonders are happened upon by aircraft, we might expect to see them starting to crop up on some maps as sites of special scenic interest—virtual tourist attractions.

And some day a new sort of symbol will be needed to mark the most historic spot: the site of the first landing by humans. Mars Society optimists are sure that day could come within ten years of the decision to go, and so in their best-case world it could be ten years from today. Pessimists argue that—if it ever happens at all—it will not happen until revolutionary technological advances in other fields (automation, perhaps, or the harnessing of fusion power) have made a radical difference to the costs involved. My own suspicion is that it will be sometime in the 2020s; failing that the 2030s are a bit more likely than the 2010s.

Putting "when" a long way off makes "where" fairly hard to guess, too. The stories we tell ourselves when looking at the maps may not be a very good guide. In post-*Viking* science fiction, early expeditions to Mars have almost all landed in the tropics of the western hemisphere. The landing sites in *Frontera* by Lewis Shiner, *Voyage* by Stephen Baxter (an oddly nostalgic book that imagines what a mid-1980s Mars mission would have been like if the ongoing program had not been canceled after Apollo), *Mars* by Ben Bova, Terry Bisson's *Voyage to the Red Planet,* Stan Robinson's *Red Mars,* Geoffrey Landis's *Mars Crossing,* and Zubrin's own novel, *First Landing,* are all within a thousand miles or so of each other.

The reason for this clustering is obvious: Valles Marineris. Its huge chasms are the most obviously exciting Martian landscape on offer, so writers converge on them. Bova's explorers head off to the canyons almost as soon as they land; Hebes Chasma is the first scientific objective explored by Robinson's crew. Landis's characters, who have gone on a Mars Direct–like mission, actually have to rappel down one side of Valles Marineris and climb up the other in order to get to a return vessel after their own ERV is lost. The great canyon even turned up in the films *Mission to Mars* and *Red Planet,* both released in 2000. The films were woeful in almost every particular, scientific, geographic, and artistic,* and had no clear or consistent view of what Mars looks like. But they both made use of vast canyons to spice up the surface scenery.

Landing near the rim of a canyon allows the explorers—or readers, depending on which way you look at it—to experience a discoverer's epiphany on first looking down. (In *Red Planet* the landing vehicle actually rolls off the plains and over a vast cliff, making the discovery particularly dramatic.) This is good storytelling but it is hardly practical. If Valles Marineris contains scientifically interesting sediments, or, as Baerbel Lucchitta and Mary Chapman have suggested, recent volcanism, it would surely be better to scout out a safe landing site as close to the science as possible, rather than set yourself down on the wrong side of a three-mile cliff. Besides, the interior of Valles Marineris has higher air pressure, which would probably make the landing easier. But for writers, space travel simply can't deliver the charge as travel across a landscape: If their characters arrived in Valles Marineris directly from orbit, without seeing it from the rim and climbing down its rifted cliffs, it would feel like cheating. And as Pat Rawlings showed in his picture *First Light,* the edge of the abyss has a special power; in narratives of exploration the emptiness beyond can stand for a planet's worth of unknowns.

The floor of Valles Marineris might be the first landing site, but so might all sorts of other places; all one can say for sure is that it's unlikely to be on one of the great dull volcanic plains such as Syrtis or Hesperia or Daedalia; it will probably be somewhere where there is either remote sensing evidence of ancient water, or of an aquifer, or of ground ice. That could be more or less anywhere, as far as we

*That said, there are degrees of woe: *Red Planet*—the one with Val Kilmer, Carrie-Anne Moss, Terence Stamp, and the firebugs weirdly referred to as nematodes—was far better than *Mission to Mars*—the one with Gary Sinise, Tim Robbins, the "Face on Mars," and the pixie-aliens from central casting. Matt Golombek was the blatantly ignored scientific adviser on *Mission to Mars* and has been stoically bearing the ridicule of his colleagues ever since.

know today. Wherever is chosen, the site will surely be commemorated with a new symbol on the maps.

It is less clear, though, that that symbol will see much further use. Barring catastrophe, it's a little hard to imagine humans never choosing to visit Mars. But it would be quite possible for humanity to visit, inspect, and not return. We can't know what precise mixture of science and symbolism will drive the first exploration in person, but the agenda may be one that can be perfectly well fulfilled with a limited series of missions, or even one. We would leave our robots there after us as we sent them on before, and maintain whatever relationship with Mars we may have come to want from a distance.

On the other hand, it may be that the first missions will be followed by the construction of permanent bases and, thus, the addition of a new symbol to the legends on the maps. Perhaps discoveries will be made that warrant protracted *in situ* inspection: a complex ancient stratigraphic system recording Mars's warm-wet past and the life it held, for example (or an Inca City, if one can be found). It's conceivable that earthly political rivalries will make it impossible for whatever powers reach Mars to quit it without being perceived as failures. Whatever the reason might be, it is quite easy to imagine Mars becoming the new Antarctica, a hard-to-reach location set aside for science under a dispensation flexible enough to permit both political rivalries, such as those of the Cold War, and political cooperation, such as that being seen today. While it would not satisfy Zubrin, something along these lines is probably what most scientists who give serious thought to a protracted human presence on Mars would want.*

With a number of bases, one might expect maps of Mars to acquire yet more new symbols, such as lines representing the transit routes that link them together. But this is not a foregone conclusion.

*The science fiction writer Brian Aldiss, an enthusiastic proponent of the Antarctic model of a Mars preserved as a "planet for science" rather than exploited for gain, at the same time suggests that its unspoiled splendors also be made available to anyone willing to pay for passage through community service as a site for "silence and meditation and honeymooning."

Rock-strewn surfaces like those seen at the three landing sites would be quite a challenge to long-distance rovers. If Mars bases have landing sites, rockets, and the energy with which to make propellants—and if they don't, their inhabitants will have a problem—then their crews may well choose to travel almost exclusively in aircraft, as they do in Antarctica. It's a usual assumption that air travel is expensive; but without a lot of travel it could work out far cheaper than putting capital into the construction of roads through the rubble. On a Mars modeled after Antarctica, we would expect people to hop from point to point in suborbital rockets; heavy equipment might be transported by means of stately dirigibles.

Long-distance surface travel would not be unheard of, but it would be restricted to journeys of experience and exploration, journeys designed to help people discover aspects of the planet—and of their lives on the planet—not captured in the ever-growing databases. People would want to take the measure of Marineris and climb Olympus, to follow the Deuteronilus shore, to cross the deep desert of Hellas. Though one suspects little will be done on foot, fat-wheeled bicycles—first suggested as ideal Martian transports in the 1950s by Robert Heinlein—might offer a way to travel while keeping in tune with the terrain. Some explorers would surely want to trek to the poles; indeed, Charles Cockell is sure they will and has even set up a foundation to encourage such endeavors (currently funding expeditions on Earth only, though). While it is an unreliable guide, it's worth noting that Mars fiction has been taking a turn to the polar recently, a reaction to the increased emphasis on ice in Mars studies and the fashion for things Shackletonian on earth.*

*Percival Lowell would not have approved. "Polar expeditions," he wrote in *Mars and Its Canals,* "exert an extreme attraction on certain minds, perhaps because they combine the maximum of hardship with the minimum of headway . . . Except for the demonstration of the polar drift-current conceived of and then verified by Nansen, very little has been added by them to our knowledge of the globe. Nor is there specific reason to suppose that what they might add would be particularly vital. Nothing out of the way is suspected of the poles beyond the simple fact of being so positioned . . . Martian polar expeditions, as undertaken by the astronomer, are the antipodes of these pleasingly perilous excursions in three important regards, which if less appealing to the gallery commend themselves to the philosopher. They involve

If technology allowed, more ambitious traverses might be attempted, perhaps over the course of a whole Martian year: a circumnavigation of the great dichotomy between north and south; a trip from pole to pole through Argyre and the Chryse trough and back again over Elysium and the Cimmeria highlands. Though Carol Stoker likes to inspire audiences with a slide that shows the route of the Lewis and Clark expedition superimposed onto the western hemisphere of Mars, these treks across Mars would not be comparable discoveries; the great reconnaissance would already be over. Nor would they become part of the geography of Mars, any more than the routes of the great nineteenth-century crossings of America, Australia, Africa, and Asia have become freeways. But they would become part of its history.

More permanent surface routes between camps and bases seem likely to be traced onto the maps only if the settlement of Mars becomes an economic affair as well as a scientific one. Zubrin has suggested that Mars might have resources of strategic minerals worth shipping to the Earth, if the costs of extraction and shipping were low enough. There is no reason to think that Mars does not have ore bodies just as the Earth does, and some might be peculiarly rich. While shipping things from Mars to the Earth would be expensive, it would be much cheaper than shipping things from the Earth to Mars. Launching a rocket from Mars is easier than launching one from the Earth and, once you're off Mars, it's downhill all the way. So if the richness of an ore body made the costs of extraction on Mars very low indeed compared with costs on Earth, there might be the basis of an export trade.

Such a trade would not necessarily lead to the creation of a surface transport infrastructure; for a big mineral strike the necessary refineries would probably be built on site, and the refined product launched straight into orbit. It's possible, though, that it might make sense to bring ore from various sites to central refineries, which

---

comparatively little hardship; they have accomplished what they set out to do; and the knowledge they have gleaned has proved fundamental to an understanding of the present physical condition of the planet." After meeting Commodore Peary he changed his opinion, a little.

might be best done across the surface, since ore is heavy stuff. And because mines would have to be built where the minerals were, they would not be able to choose the sorts of sites where self-sufficiency is possible. A mine in a ground-ice-free zone might need shipments of water from the north or south. A mine in the Prometheus basin close to the south pole would need food shipments during the months of darkness when its greenhouses would have to close down.

If, for these reasons or others, things need to be moved in bulk, then one might imagine roads or railways starting to spread across the map. For various reasons, science fiction has concentrated on railways; they turn up in all sorts of Mars fiction as the natural way to move from place to place, and sometimes even become the focus of the story. In part, this is just another facet of the Mars-as-the-West trope; the railway is a defining if paradoxical symbol of the frontier West, since it is with the coming of the railways that the frontier becomes accessible and ceases to be the inspiration it once was. The railway fixation is also part of science fiction's more general tendency to re-create the infrastructure of the early twentieth century. To make Martian land transport railwaylike is part of the same approach to the universe that makes spacecraft ocean linerlike. (This must have been a very natural approach for early American science fiction writers, many of whom were second-generation immigrants with a clear understanding of what sort of transport got you to a new world.) Railways lend themselves to shared experience more obviously than private automobiles, and they appeal to the machine-minded in a more thoroughgoing way. They turn a landscape and the society that inhabits it into something comprehensively mechanical, surveyed and synchronized; it was a nineteenth-century commonplace that the railway network as a whole was a single machine, the individual trains components of something greater, like shuttles in a loom. The railway abolished distance, as the Internet is said to do today—and the abolition of distance is a prerequisite for certain forms of escapism, even though, in the end, it makes true escape less feasible than ever.

There may come a day when railways replace rockets as a way of getting around the surface of Mars. There may also come a day when railways, of a specialized sort, start to compete as ways of getting off the planet. Rockets are essentially wasteful ways of getting into space, because they expend so much more energy in lifting themselves and their fuel than they do in lifting their cargo. More efficient launch systems would accelerate the payload up to orbital speeds while keeping the heavy propulsive hardware on the ground. In the 1860s Jules Verne imagined firing men to the moon from a vast cannon; in the 1950s Arthur C. Clarke came up with the rather more practical idea of using an electromagnetic catapult based on similar principles to those used in magnetic levitation trains. The amount of acceleration required would depend on the length of the catapult; if it could be made long enough, the acceleration might be kept low enough to avoid killing the passengers, which is normally seen as a plus in railway management. This approach might be well suited to Mars; among other things, the ideal site for such a catapult is on a steady upward slope at the equator, and Mars has just the thing at the top of the Tharsis rise. An electromagnetic railway track that accelerated its cargo at two Earth gravities across a couple of hundred miles of Tharsis and up the gentle western incline of Pavonis Mons would be able to throw things right into orbit. If such a catapult significantly undercut the costs of rocketry it would make sense to ship all the planet's exports overland to Pavonis, and Mars might have the railways its chroniclers have always wanted.

Another even more ambitious idea was thought up by Russian engineers and American oceanographers in the 1960s. Imagine a satellite in synchronous orbit around a planet—that is, in an orbit that takes it round the planet in exactly the same time as it takes the planet to rotate, so that the satellite seems to stay fixed over a particular spot on the equator. Now imagine reeling out two very long cables from this satellite, one down to the planet, one out toward the stars. (You need two cables to keep the whole thing balanced.) When the cable going down hits the surface, wrap it around a very strong cleat. You now have a physical bridge into outer space—a "cosmic funicular," as some of the idea's Russian pioneers put it.

Instead of throwing things into orbit you could just drive them up the funicular.

The problem with such space elevators is that they must be able to support their own weight, so building them requires materials of incredible strength, such as single crystals of carbon, in prodigious amounts. These constraints put a space elevator for the Earth at the very edges of the conceivable. As Arthur C. Clarke pointed out in his novel *The Fountains of Paradise,* though, a space elevator for Mars is comparatively easy. A synchronous satellite around Mars is in a lower orbit than one around the Earth, and the planet's gravity (and thus the weight of the elevator) is lower. The biggest problem, as Clarke realized, is that the elevator would cross the orbit of Phobos, the inner moon. But this problem is not insurmountable. The space elevator, like a taut string, would have a natural tendency to oscillate, and it should be possible to tune those oscillations in such a way as to avoid Phobos. I don't mean to belittle the problem of taking a twenty-five-thousand-mile-long structure made of millions of tons of something like artificial diamond and tuning it to harmonize with one of the hurtling moons of Mars. But any engineers in a position to build such a thing in the first place could surely take such a detail in their stride.

If Phobos offers the beanstalk builders a bit of a challenge, though, Pavonis Mons again offers some help. Its summit, right on the equator and well above the worst of the weather (it stuck out of the great storm of 1971), is a natural ground floor for a space elevator. If you built the elevator anywhere else, the vast asymmetric mass of Tharsis would endlessly be pulling it off balance; build straight down to the center of Tharsis and this problem would be minimized. So if the Martian economy ever grows large enough to justify the building of either a catapult or a space elevator, we can be pretty sure that Pavonis will come into its own as the doorstep to orbit and earn itself a pretty special symbol on the maps. It is one of those rare cases where geography is destiny and part of that destiny would be as the hub of the planet's railway system. The railway lines, we can be reasonably sure, would be straight and dramatic. Indeed, they might look not unlike Lowell's old maps. As some of Lowell's critics pointed out, his straight lines were an unlikely arrangement for irrigation canals, since canals tend to follow contours of constant elevation. Railways, on the other hand, cut through contours with zest. Engineers with access to materials strong and light enough for a space elevator would be able to lift their lines over any obstacles with ease; in the Martian gravity one can imagine delicate bridges leaping the narrower parts of Valles Marineris in a single span.

A Martian railway network could be a work of art on a planetary scale. But it might not be the only one. The Japanese-American sculptor Isamu Noguchi, perhaps most famous today for his furniture design, worked in his youth with Gutzon Borglum, the creator of Mount Rushmore; concerned with the possibility of planetary death after Hiroshima, Noguchi designed a stylized face several miles long called "Sculpture to be seen from Mars." Though never built, it was, like some of Noguchi's other grand ideas, a precursor to later attempts to see the landscape itself as something for artists to mold and shape, an idea that has since taken root in the American West in the form of strange earthworks, unnecessary jetties, fields of lightning rods, and James Turrell's project to reshape Roden Crater, north of Flagstaff.

This is a peculiarly easy school of art to imagine on Mars, an art of simplicity in the wilderness, an art that seeks to evoke the planetary perspective. A single trench in the emptiness; a crater wrapped in foil, Christo-style; rocks formed into a drystone wall, or a circle of standing stones*: Any intervention could be art. How about a version of Noguchi's face—a sculpture to be seen on Mars, not from Mars? Such art would not necessarily mean moving vast amounts of regolith (though it might). In 1980 the artist Tom Van Sant made the largest artwork in history with twenty-four two-foot-square mirrors. They were arranged in the desert north of Los Angeles (not far from the Goldstone antenna) in such a way that they would reflect sunlight up onto the camera of a satellite passing overhead. The result was the image of an eye a mile and a half long burned into the satellite's detectors pixel by pixel, a work of art as

*In both Robinson's *Red Mars* and Bill Hartmann's novel *Mars Undergound* Martian artworks mimic Stonehenge.

transitory as it was vast, seen only for a fraction of a second by a single robot but recorded for all time.

If neither railway commerce nor avant-garde art rearrange the surface of Mars, though, scientists—the first Martians—might do the job themselves. Of all the experiments that might be performed on Mars in the next century or so, perhaps the most interesting and almost certainly the most spectacular would be to create a new crater. Not a little nuclear crater like Teapot Ess or Jangle U, but a decent crater a few miles across, bigger than Meteor Crater, smaller than Haughton. There would be various reasons to undertake such a project. One is that, in the long run, the ability to divert asteroids and even comets will become a near necessity regardless of what goes on on Mars. A significant asteroid impact can be expected on the Earth every couple of centuries and a truly cataclysmic one—one a fair fraction of the size of the Haughton crater and able to kill a season's crops worldwide—perhaps every two hundred thousand years. To avert such catastrophes, humanity will need the ability to change the orbits of asteroids, and this is a skill that it would probably be best to practice in advance.

If one starts from the assumption that it will be necessary to push some asteroids around, then the idea of putting one on a course for Mars is natural. While a lot is already known about the processes of impact cratering, no one has ever actually seen a large crater being formed and such observations would be wonderfully instructive. And creating a crater is not a purely destructive act; it might be environmentally benign. One of Charles Cockell's objectives of the Haughton Mars Project is to try to understand the high-temperature microbial ecosystems that colonize impact craters in the immediate aftermath of their formation, when the rock within and around them is as hot as a volcano. If there is any microbial life deep under the surface of Mars—or frozen, dormant, in its permafrost—it's a fair bet that it knows how to take advantage of the pulse of warmth and melted water that undoubtedly follows all large impacts. Impact aftermaths may be transitory Martian oases.

The addition of a man-made crater or two would make a change to the Martian maps that went beyond the symbolic to the topographic. Other experiments might go still further. It's not inconceivable that a human presence on Mars might eventually start redrawing the basic physical parameters of the map: flooding craters, shrinking ice caps, even re-creating some of Tim Parker's shorelines. A human presence could go beyond making Mars a new world. It could start to turn it into a new planet.

Gaia's Neighbor

Man is one world, and hath
Another to attend him.

—George Herbert, "Man"

In the 1960s Jim Lovelock, an English scientist, found himself spending quite a lot of time in Pasadena. Lovelock's claim to fame was an ability to invent chemical detectors of immense sensitivity; indeed, his detectors' unprecedented sensitivity was changing the world. It was Lovelock's electron-capture detector that picked up traces of organic pesticides throughout the food chain, thus providing much of the spur for Rachel Carson's *Silent Spring*. In the 1970s his technology would reveal how widespread the chlorofluorocarbons (CFCs) used in aerosols and refrigerators had become in the environment at large and thus trigger the global debate on the ozone layer. JPL wanted to use similar instruments to study the environments of other planets, and Lovelock, who had taken the bold and unusual step of leaving an enviable position at a world-class laboratory to become a freelance researcher, was happy to be a consultant to them.

In 1964, Lovelock went to a meeting about instruments that might look for life on Mars. Vance Oyama described ways of looking for chemicals typically produced by living organisms; Wolf Vishniac discussed ways to try to encourage measurable growth in microorganisms by feeding them certain nutrients. (On Earth this method—called the "Wolf trap"—proved highly sensitive; on the Antarctic trip that would end in his death, Vishniac proved the existence of life in places previous studies had proclaimed sterile.) Lovelock was not impressed. These approaches would detect Martian life only if the lander chanced upon soil that contained living creatures and if those living creatures had metabolisms like those of earthly microorganisms. A true test for life on another planet, Lovelock thought, should be one that looked at the planet as a whole, rather than a particular place, and one that looked for some sort of marker that would necessarily be common to all conceivable forms of life.

Guided by these ideas, Lovelock came up with a different approach to the search for life. Schrödinger had suggested that the fundamental property of life was to reverse entropy—to impose local order on a universe that was, in general, running down into chaos. The particular type of order that Lovelock decided to look at was chemical (he was trained as a chemist) and planetary: the make-up of the atmosphere. On the Earth, he argued, life's presence is clearly seen in the atmosphere, which life uses as both a source of raw materials and as a place to dump its waste. The fact that life keeps using the atmosphere in both these ways means that it is out of balance, containing a mixture of chemicals that can react with each other. Methane and oxygen, for example, react together quite vigorously (which is why Zubrin's rockets use them as propellants), and yet both are present in the Earth's atmosphere. This is only possible because life is endlessly replenishing the atmosphere's reactive gases; without life the atmosphere would quickly reach chemical equilibrium, the state in which every chemical reaction that could make a difference has already done so. Chemical equilibrium means maximal entropy; to maintain a state far from equilibrium means entropy must be actively reduced.

Lovelock decided that a sufficiently off-balance atmosphere would constitute clear evidence for an entropy reduction caused by life. It didn't have to be off balance in the same way that the Earth's atmosphere was; it just had to contain chemicals that, left to themselves, would react together. Soon after, earth-based spectroscopy confirmed that the Martian atmosphere was as close to chemical equilibrium as makes no matter. To Lovelock the case was more or less proven: Mars was dead and the biology experiments being discussed at JPL—the experiments that would end up on the *Viking* landers—were more or less pointless. This was not what JPL wanted to hear. But though the experimental results were not absolutely clear-cut, the post-*Viking* consensus ended up exactly as Lovelock had predicted: There was no life on Mars.*

If Jim Lovelock's insights did little to change the way JPL approached Mars, they did a lot to change the way he himself looked at the Earth. The Earth's atmosphere is not just far from chemical equilibrium; it is also, bizarrely, pretty stable. Oxygen levels, for example, have changed only a little over the past 250 million years. Lovelock came to the conclusion that this, too, was evidence of life—that living things somehow acted to regulate the instabilities they caused in their environment. Life and its physical environment, he came to believe, were not separate entities to be dealt with by biology or geology, as the case might be; they were part of a single self-regulating system. Following the advice of his neighbor, the novelist William Golding, Lovelock took to calling that self-regulating system Gaia, after the ancient Greek goddess of the earth.

In the 1970s, Lovelock's ideas were by and large rejected by the scientific establishment while eagerly received by the green New Age fringe; neither group really understood what he was saying. But Boulder's original Mars Underground, somewhere between the two, grokked him in fullness.† Gaia was not just a theory about the Earth; it was a theory about life as a planetary phenomenon. And as such it could be applied to Mars.

The question of life on Mars raises problems on at least two counts. The first is that a longing to answer it in the positive encourages people to mislead themselves on the subject. The second is that, if you see the search for life as a search for individual living organisms, it is almost impossible to answer in the negative; to say categorically that there is no living thing on Mars would require knowing the planet with an intimacy that is hard to imagine, especially now that the deep subsurface has been brought into play.

*The hypothesis of life deep below the surface does not necessarily contradict Lovelock's ideas, inasmuch as he explicitly assumed that life would be in a position to make continuous use of the atmosphere, which would not be the case for a deep biosphere.
†In the Martian religion central to Robert Heinlein's *Stranger in a Strange Land* to grok, literally to drink, is a signifier of understanding and oneness.

How long do you have to look for fossils before you decide there can't be any? If you do find fossils, how long must you scour the ice to be sure that not a single viable bacterial spore survives?

An alternative, in many ways more illuminating, approach is to look for possibilities rather than facts. Instead of asking whether there is life on Mars, ask whether there could be. This opens up new routes of inquiry. It focuses the mind on the many complex links between life and its environment. And it is not limited to the past and present. It is a question that applies to the future.

By 1976, this mostly idle notion had a name—terraforming—and a history, most of it in science fiction, but some of it in articles in reputable journals. There had even been a small workshop on the subject at NASA Ames. In Boulder, the Underground seized on the idea. Faced with a huge amount of information about a seemingly lifeless planet, it was a topic that appealed to their scientific romanticism more than any other (except for the dream of actually going there). Terraforming was the scientific story of the planet's decline translated into technology and told in reverse. It was not all that they were interested in—far from it—but it helped to give structure to everything else. And much of that structure came from Lovelock's ideas, which by the late 1970s had been synthesized in his book *Gaia: A New Look at Life on Earth.* Traditional terraforming, if we can speak of such a thing, was largely about physical alterations of the environment that would make life possible: The first novel to deal with terraforming Mars, Arthur C. Clarke's *The Sands of Mars,* imagined using nuclear power to turn Phobos into a second sun. Lovelock's vision led Chris McKay, Penny Boston, and the rest of the Boulder gang to distrust such pure engineering ideas as too simplistic. Terraforming should not just be about creating a new environment for life through *force majeure,* but about finding ways to allow life to create a new environment for itself.

These ideas were helped along by the belief that in some respects terraforming Mars might be quite easy. Carl Sagan's idea that Mars's orbital cycles might create "long winters" suggested that the climate had two distinct states: The one we see in which much of the atmosphere is frozen into the ice caps and the soil; and one in which the atmosphere was freed to warm the surface through the greenhouse effect. If the terraformers could switch the climate from one natural state to the other, then they'd be well on their way. To make things easier, the switch was weighted in the terraformers' favor; given a gentle nudge, it would snap itself all the way on. If you could just get a little extra carbon dioxide into the air, you'd warm up the planet enough to get a bit more; that would warm the planet yet further and get you a bit more; and so on. The greenhouse effect would amplify itself by coaxing more and more carbon dioxide from the caps and the frozen soil. For evidence that such a thing was possible its proponents could point to Venus, which owes its intolerably hot surface to just such a runaway greenhouse effect. On Mars, where solar radiation is only a quarter what it is on Venus, intolerable heat was not going to be a problem, but less bitter cold was a definite possibility.

Warming the planet's carbon dioxide reservoirs a little might thus trigger a process that warmed the whole thing a lot. The most obvious way to do this was to add some soot or dust to the polar caps, making them less reflective. Or you could put mirrors into orbit to focus extra sunlight on them. Or, taking a more Gaian approach, you might be able to design a dark algal crust that would spread over them. By the time of the first public colloquium on terraforming, organized by James Oberg* and held as an evening satellite meeting at the 1979 Lunar and Planetary Science meeting in Houston, some form of polar melting was seen as the established first step in terraforming Mars, despite the fact that Sagan's ideas about natural "long winters" were by that time out of favor, having been largely replaced by the far slower, noncyclic climate change championed by Jim Pollack. At the most recent major discussion of terraforming, held at NASA Ames in late 2000—a meeting attended by a surprising number of the people who had met in Houston more than twenty years before—getting carbon dioxide out of the caps and soil and into the atmosphere was still seen as the obvious first step.

The positive feedback in such an approach—the fact that enough energy to heat up the planet by 1°F can, thanks to the greenhouse effect, actually produce 12°F of warming—definitely makes it powerful. But changing a planet's climate still takes quite a lot of effort.

*Who also wrote the first book on the topic, *New Earths.*

Sagan calculated that you might be able to release enough carbon dioxide to kick the process off by covering just 6 percent of the caps with just a twenty-fifth of an inch of soot. This may not sound like much, but it still means moving a hundred million tons of material. And since you wouldn't have soot available, you'd need to use powdered basalt, a less efficient heat absorber, and cover more of the cap. And a layer a twenty-fifth of an inch thick would blow away in the wind; you might want ten or a hundred times more of the stuff. Even this simple modification requires pulverizing and redistributing billions of tons of rock.

Mirrors might be simpler. Instead of increasing the efficiency with which the cap absorbs sunlight, mirrors poised in orbit above it could simply increase the amount of sunlight it receives. According to calculations made by Robert Zubrin and Chris McKay in the early 1990s, a space-based mirror 150 miles or so across would do the job nicely. Made from very thin foil, such a mirror might have a mass of two hundred thousand tons, which compares well with the billions of tons of crap needed to darken the cap. At two hundred thousand tons, such a mirror would be only half the mass of the largest oil tankers. What's more, it could be delivered in self propelling installments. Light bouncing off a mirror exerts a small pressure as it does so; a very thin foil mirror could literally sail on sunlight, navigating the solar system without using any fuel at all. One can imagine sail after sail being produced at a smelter on an aluminum-rich asteroid and tacking through the sunshine to Mars, where they could use the pressure of the sun to balance the force of gravity and hover steadily a half a million or so miles away from the planet, reflecting sunlight to the polar cap. It's not necessarily the case that such industrial capabilities will ever be developed in space, but it's not impossible. And flying in two hundred thousand tons of foil from the asteroid belt compares pretty well with moving billions of tons of dirt over the surface of the planet.*

*However, there is another, nontechnical issue here that might throw Zubrin's ideas off track. If asteroid mining is advanced enough to produce fleets of orbiting mirrors for a terraforming project, it might be a serious competitor to any mining on Mars and thus reduce the need for a Martian frontier. Zubrin's answer—he always has one—is that if miners go to the asteroids, Mars can get rich by supplying them with food and other replenishables much more cheaply than the Earth could.

A further possibility was raised in the 1980s by Jim Lovelock himself, drawing on his own experience. On the Earth, CFCs have come to be seen as a problem both because they deplete the stratospheric ozone layer and because they trap infrared radiation that is leaving the Earth far more effectively than more prevalent greenhouse gases like carbon dioxide do. On Mars, Lovelock realized, these greenhousing powers might be just what are needed. Concentrations of CFCs as low as a few parts per billion might warm the planet enough to kick off the desired runaway greenhouse effect. His tongue in his cheek, Lovelock suggested shipping the CFCs not wanted on the Earth over to Mars as soon as possible.

When Chris McKay started to analyze this idea, it did not look promising. For a start, even a gas present in the atmosphere at levels of only a few parts per billion is still there in large quantities, since an atmosphere, even a thin one, is a very big thing. Shipping from the Earth would be impossible. What's worse, the CFCs would be broken down by ultraviolet light; the only reason they last long enough on the Earth to percolate up to the stratosphere is that the ozone layer they attack when they get there protects them on the way up. On Mars, hard ultraviolet rays would start destroying the CFCs as soon as they left the factory chimneys, which would mean the factories had to labor even harder in order to keep a satisfactory CFC greenhouse in place. According to calculations McKay and his colleagues made for a terraforming article that the prestigious journal *Nature* published in 1991, a CFC greenhouse would require an industrial base capable of producing a trillion tons of gas a year. That's something like a million times the amount that was produced on the Earth when the use of CFCs was at its peak.

When McKay did the calculations in the early 1990s, this vast level of production made the supergreenhouse effect look unattractive even by the exuberant standards of techno-maximalist terraformers. By the time of the 2000 meeting at Ames, though, things were looking up. Gases that were far better infrared absorbers than everyday CFCs had been discovered, and theory suggested that even better supergreenhouse gases might be possible. (Unsurprisingly, not much research goes into making better greenhouse gases. But a handful of theorists find such ideas interesting.) Better still, these new gases were much less susceptible to UV, and might last in the Martian atmosphere for thousands of years. It was possible that a remarkably effective supergreenhouse could be maintained through the production of just a few hundred thousand tons of supergreenhouse gas every year. There would be no need for the mass production of solar sails at asteroid factories; a dozen chemical plants and a few large mines would do fine. (A facility on the scale of South Africa's Vergenoeg mine, a large fluorspar extraction operation, could produce the raw material for about fifty thousand tons of gas a year.) If Mars has fluorine resources comparable to the Earth's, it would be possible to keep this up for thousands of years; geochemists suspect Mars may in fact be significantly richer in fluorine than the Earth is.

In the long run, such factories and mines might not be necessary. It is conceivable that such gases might eventually be synthesized by plants and bacteria. The genetic engineering involved would be heroic, as it would involve the design of entirely new metabolic pathways, but it is not unimaginable. So it might be possible to create a fluorine-bearing-gas cycle on Mars that would naturally maintain the greenhouse: When the gases were broken down in the atmosphere, living creatures would metabolize their remains and create fresh supplies of the stuff. A few hundred thousand tons a year, while still quite a lot by industrial standards, is relatively small change by the standards of a living planet. One of Lovelock's early Gaian insights was that oceanic plankton pump sulfur out of the water and into the atmosphere, thus making the stuff available for creatures on the land; they do this by releasing about seventy million tons of dimethyl sulfide every year. Martian plankton would have to produce less than a tenth that much supergreenhouse gas to keep their planet warm.

One way or another, triggering a runaway carbon-dioxide greenhouse on Mars, and thus providing enough warmth and air pressure to allow liquid water on the surface for a significant part of the year, seems to be technically feasible. It's not possible today, nor will it necessarily be possible this century, but if humanity's technological prowess continues to grow, and if the technologies of spaceflight are part of that growth, at some point this first stage of warming will become a practical proposition. The question then will be whether to make it happen—and what to do next.

This is one of the central questions of Kim Stanley Robinson's epic Mars books. Some of his settlers want to terraform their planet as quickly as possible; others want to leave it as it is. In a nice play on earthly politics, the groups become known as Greens and Reds respectively. The logic of the novels requires the Greens to triumph, but the Red case is taken seriously and portrayed with sympathy. Robinson is a man who prefers bare granite to soft limestone. He knows that the surface of Mars is a vast natural sculpture, a billion-year symphony of slow stone. Barren, yes—and beautiful not so much despite that as because of it. Warm it and you deface it. Melting permafrost causes crater rims and ridges in the high latitudes to slump undefined to the even plains; opened aquifers undermine the ground itself, causing its chaotic collapse. Once moisture gets into the air the dust turns into mud all over the planet. The mud slides. Clear skies turn to cloud; wind-carved rocks are eaten up by acid rain (with that much carbon dioxide in the atmosphere, all rain will be acidic). Newly thickened storms blow away the delicate strata of the terraced terrains; dirty fizzy waters lap up over the sand dunes of the northern erg. What was once simple and grand becomes messy and muddy.

While Robinson faces up to much of this, he avoids a couple of key issues. He makes the terraforming far faster than is conceivable given current knowledge, producing an atmosphere that humans can breathe within a century or two. And he ignores the question of native life. His books, conceived in the 1980s, make the reasonable simplifying assumption that there is no native life on Mars and never has been. In reality, though, that case is not proven.

To some it doesn't matter. When he first started talking about terraforming in the early 1990s, Zubrin seemed to take some delight in not caring about what effect, if any, the process might have on indigenous life. After all, indigenous Martians, if there are any, would almost certainly just be bacteria—and anaerobic bacteria to boot, creatures humans associate only with decay, at home only in the cesspit and the gangrenous wound. As Zubrin pointed out, we kill such creatures in their millions with bleach and antibiotics whenever they discommode us. If similar creatures are to be found on Mars, then study them by all means—but don't let their existence alter your plans to change the planet.

When Zubrin was a brilliant loudmouth speaking only for himself, this contentious view was not much of a problem. Once he was founding president of the Mars Society, though, it became an issue. The Mars Society is an undeniably utopian and escapist organization, and there is nothing wrong with that. As J. R. R. Tolkien once remarked, to be opposed to escape is to put oneself on the side of the jailers. What the society's members think they are escaping, though, is a matter on which opinions differ. Zubrin wants to escape the stifled world that lacks a frontier for a utopia of human victory over natural obstacles. But many of those who came to the founding convention of the Mars Society wanted to cherish the environment, not overcome it. Where Zubrin wanted to restart the history of the frontier, they wanted to rewrite it—and to do it better this time. They looked back at frontier America with mixed feelings, delighting in the land that was revealed and passed on to them but dismayed at the human and environmental costs of the expansion. Zubrin's damn-the-torpedoes disregard for Martian life was exactly the sort of attitude they wanted to escape; to the extent they wanted to terraform at all, they saw the act as a restitution, even an atonement. They didn't want a new America; they wanted a dream of America done differently. They wanted a Martian wilderness wherein might lie the preservation of the human world.

The war of these world-views was played out on the evening of the second day of the Mars Society's first convention. The terraforming panel was dominated by people with views as robust as Zubrin's or even more so: The general tone was that Mars was there to be taken and terraforming was the way to take it. The voices from the floor were angry. Mars was something wonderful, something worthy of respect, not just a means to an end—and if it had lifeforms, even lowly ones, they were worthy of respect and wonder, too. When the panel talked of manifest destiny, or indeed genetic destiny, the discontents in the audience talked of the slaughter of the native Americans. Such talk got under Zubrin's skin: To equate humans with Martian bacteria was madness. It was offensive and, almost as bad if not worse, it missed the point. Going to Mars, to Zubrin, is not really about Mars. It's about the human race doing what it does best.

His antagonists, though, are as interested in the environment as in the people, or at least try to be. Look at a painting of the Americas as wilderness by Cole or Church and you see the sort of thing they yearn for: unspoiled land from which to take solace. They see such wilderness not as a resource, but as a source—the source of the life that they themselves share. They want a life at one with nature, and they ignore—possibly perversely, and possibly with a great, unacknowledged insight into the nature of creation—a basic truth about Mars: that it looks like a place where nature and life are separated. Life may simply not be in the nature of Mars; the nature of Mars may be to be barren.

This is why the Martian environment and its possible protection make such a fascinating topic for discussion. In offering nature without life, Mars reveals itself as a precise complement to what may be the major cultural shift of the twenty-first century: The realization that the living and the natural are not necessarily the same. Biotechnology throws the relationship between life and nature into question. As biotechnological abilities progress, we will be forced to recognize that living things are not necessarily natural, and to disentangle the respect we feel for nature and the respect we feel for life. Thinking about terraforming brings the distinction into a peculiarly clear focus. To honor and value life on Mars is not to value nature, at least not if Martian nature is lifeless. To spread life on Mars would be an act of destruction as well as of creation. This insight will not lay the issue to rest; but it may help us think about it better. And for all the appeal of the great empty desert in the sky, it may help us to see that, in the end, it is life that we value most highly.

Though such ideas may come to have a growing cultural resonance, in the Mars Society at least they are no longer proving divisive. The rhetoric has been toned down: The term "*Lebensraum*" was only ever used on the fringe (and then much to Zubrin's distaste) and is not heard today; Stan Robinson, speaking at the Society's second convention, argued persuasively that "Manifest Destiny," too, should be dropped from the lexicon on the simple four-word basis that "it reeks of murder." The differences of opinion remain—how could they not, in an organization with such different dreams that Robinson can write novels about a Mars with little or no private property while Zubrin wonders how to get legal authority to sell off land rights—but have been used to define a latitudinarianism in the Society, a quality a collection of highly focused engineering types sorely needs. The members have found ways respectfully to disagree on the issue. At that first convention, though, the debate was not just rowdy; it was rancorous, and it lasted long into the night.

Chris McKay, who had the unhappy task of trying to moderate that debate, had undoubtedly thought longer and harder about the issues than anyone on the panel. In the late 1980s he'd started writing papers that looked at the ethics of terraforming, as well as the technologies. He'd asked himself what rights an ecosystem had: What were the rights of life, and what were the rights of rocks that had no life? And he'd wondered what difference the provenance of the life on Mars might make. If there are Martian microbes still alive in the depths—which, despite having been one of the authors of the first paper to discuss the subject, McKay doubts—where did they come from?

The problem with life on Earth is that for all its bewildering variety it is in some respects all the same. Every living thing on Earth shares a common ancestor; every living thing uses the same basic biochemistry of nucleic and amino acids. It is thus very difficult to distinguish between those attributes of life that must be common to all living things everywhere and those attributes that are simply aspects of the way that life has evolved on Earth. This is one of the reasons why the idea of finding life beyond the Earth is scientifically fascinating; until you have two different examples of what life can be like, you're very limited in your understanding of what exactly it is and of how it can get started in the first place.

Discovering life on Mars might provide a way to answer those questions. But it might not. One thing that ALH 84001 and the rest of the Martian meteorites prove beyond doubt is that rocks can travel between the surfaces of planets; the theory is that the shock wave from a large asteroid impact sweeps them up and shoots them off like a primitive and rather rougher version of a launching catapult. There is no reason to believe that this process would sterilize the rocks launched into space. Space itself will not necessarily sterilize them, either. *Apollo 12* landed near an earlier automated moon probe, *Surveyor 3*, and took samples from it back to Earth. Some of these samples proved to have viable bacteria on them, hitchhikers not notably incommoded by their trip to the moon and back. The Long Duration Exposure Facility, a NASA satellite that was put into orbit by the shuttle *Challenger* in 1984 and recovered by the *Columbia* in 1990 (a rather longer duration of exposure than intended, due to the *Challenger*'s subsequent accident), also returned to Earth with its bacterial entourage intact. And bacteria do not need an Apollo mission or a space shuttle to bring them down to Earth safely; careful analysis of ALH 84001 by Joe Kirschvink shows that most of the meteorite came through the process of entry into the Earth's atmosphere without being heated up to anything like the temperature needed to pasteurize the passengers. (Meteorites heat the atmosphere as they slough their surface layers off in it, which is why we see them as shooting stars, but what remains when they hit the ground is not necessarily warmed up in the process.)

Martian meteorites are not common, but nor are they extraordinarily rare. Estimates suggest that there are a million or so orbiting the sun at the moment, and over time many of them will hit the Earth. In the early solar system, when there were many more asteroid and comet impacts to launch the things, Mars rocks would have been far more numerous. And although it is harder to launch rocks from the Earth than it is from Mars, there would have been a steady stream of meteorites headed in the opposite direction, too. Mars and the Earth, as McKay likes to put it, have been swapping spit for billions of years.

There has been life on Earth for at least 3.8 billion years. It is inconceivable that in that time no rocks carrying earthly bacteria have made the journey to Mars. While it may be that most of those that made the journey were sterilized by vacuum and radiation—the fact that bacteria can survive a few years in space does not guarantee that they will survive a few million years—some will have reached Mars relatively quickly, with the bacteria on board still viable. And it is quite plausible that, if they found conditions on Mars to their liking, those bacteria made themselves at home. If living creatures or fossils are ever found on Mars, the possibility that they are related to life on Earth will have to be investigated. If all there is to work with is a bunch of fossils, the question may never be answered. But if there are living samples—or dead samples that have been preserved in deep ice for billions of years, rather than turned to rock—molecular biology will come to our aid. If the "Martians" use the same genetic code that earthlife does, then it will be a racing certainty that they are our cousins most removed.

Indeed, it would be quite possible that they, not we, represent the ancestral branch of the family tree. The basalt in ALH 84001 predates the great collision that threw the makings of the moon into orbit round the Earth. Regardless of whether ALH 84001 carries signs of life, it shows that Mars has had a solid surface for longer than the Earth has. And if Mars has relatively little water—enough for a small northern ocean but not much more—that may have helped keep it habitable throughout the solar system's boisterous youth. Really big impacts on a water-rich early Earth would have boiled the oceans and produced a temporary atmosphere of pure steam. Since steam is opaque, it takes a long time to cool down, because the heat within it can't be radiated away. The steam-bath atmosphere would thus persist for a couple of thousand years and the planet's surface would be thoroughly sterilized down to a considerable depth. If Mars were less well endowed with water its surface would never be steamed in quite the same way. So to some Mars looks like a more likely site for the origin of life than the Earth. The ever-inventive Norm Sleep, of Stanford, has suggested that a Martian origin might explain why life on Earth is separated into three distinct lineages, the prokarya and archaea (all single-celled) and the eukarya (animals, plants, fungi, and all sorts of other good stuff). The three great families might represent three distinct sowings of Martian seed across the interplanetary void.

For Chris McKay these events in the past have a crucial impact on the morality of the future. If there are no Martians, planetary engineering raises few problems; a world with life represents an unconditional improvement on one without. If the Martians are our longest-lost cousins the situation is similar; we share the Earth with all sorts of strange distant relatives and we can do the same on Mars.

But if the Martians are not related to us, things are different. We would not be justified, he thinks, in exterminating the only truly alien life we know, nor in appropriating its environment to our own ends. But that does not mean we should do nothing. After all, any Martians around today are hardly pulling their weight as good Gaians: They show no evidence of keeping an intrinsically unstable environment stable, of being part of a self-regulating system. Perhaps they need a helping hand. Providing them with a warm, wet surface might be just the new opportunity that they needed—as long as we leave them alone to make of it what they will.

At this point we are talking about something other than terraforming as the term is usually understood. The idea is no longer to make something like the Earth; it is to make something new, or re-create something old, with only passing reference to the Earth. The late Canadian biologist Robert Haynes—whose name was given to Haynes Ridge at Haughton Crater, the site of the Mars Society's Flashline hab—coined a more general word for such an undertaking: ecopoiesis, from the Greek roots meaning abode (eco, as in economics or ecology) and fabrication (poiesis, as in poetry). It is quite possible that, if there is life on Mars, a limited ecopoiesis aimed at allowing that life to spread back to the surface would be more acceptable to public opinion than true terraforming aimed at making that surface habitable to humans. It is certain that the ecopoiesis would be much easier. For such an ecopoiesis the relatively quick and gentle warming and dampening of the planet that might be brought about by triggering a runaway carbon-dioxide greenhouse would probably be enough, and might be achieved in a couple of centuries. Terraforming proper—making Mars an environment where earthlife can prosper—is a much harder proposition. Ideally it requires an annual average temperature above 32°F, a relatively copious supply of water at the surface, and an atmosphere of oxygen and nitrogen.

The temperature might be the simplest thing: After initial warming with a runaway greenhouse, the temperature could probably be regulated reasonably well using supergreenhouse gases. If there were enough carbon dioxide released by the original warming, the oxygen could be produced through photosynthesis, if you had hardy enough plants and were willing to wait for a few thousand years. But there's a drawback to photosynthesis: It creates organic carbon compounds—leaves, wood, and the like—as well as oxygen, and those would have to be disposed of. Otherwise bacteria would just use up the oxygen while eating dead plants, reproducing the carbon dioxide you'd started with as they did so.

Releasing an ocean frozen into the megaregolith would be a help here, because oceans are great places for burying organic carbon. The problem is that releasing such an ocean would be a titanic undertaking; melting ice caps is child's play by comparison. Simply heating the surface and waiting is not an option: Because the rocks are cold and heat takes a long time to diffuse, it would take more than a million years for the top two miles of the planet to thaw out, and a large part of the water is probably a fair bit deeper than that. The only way of melting that water that we can easily imagine is through the use of nuclear explosives in numbers that would make Dr. Strangelove quail—vast fallout-free hydrogen bombs made from Mars's reasonably copious supplies of deuterium. According to Martyn Fogg's authoritative book *Terraforming: Engineering Planetary Environments,* something between a hundred million megatons and ten billion megatons of nuclear explosives would be required to release a serious ocean if it were frozen solid (if it were liquid under a cap of ice, things get much easier).*

While we're on the subject, nuclear explosives might also help with the nitrogen. The Martian atmosphere has very low levels of nitrogen, but it is conceivable, even likely, that there is a good bit more stowed away as solid nitrates in the crust. There are bacteria that turn nitrates into pure nitrogen; they do it constantly on the Earth, producing three hundred million tons a year of it, mostly from decaying organic matter. Unfortunately, at that rate it would take over half a million years to produce enough nitrogen for a vaguely earthlike atmosphere on Mars. If there are concentrated beds of nitrate, nuking them might get the stuff out quicker and

*Ralph Lorenz, a planetary scientist who works with Peter Smith at the University of Arizona, has pointed out that the tides raised in such an ocean would significantly speed up the decay of Phobos's orbit, hastening the moon's destruction by many millions of years. Long-term planners, take note.

release some helpful oxygen, too. But if the nitrates are spread all over the planet this becomes difficult. It would probably be simpler to ship in the nitrogen from somewhere where it's plentiful, such as Saturn's moon Titan, which has an atmosphere full of it. If, that is, you can find a way to move trillions of tons of gas billions of miles across the solar system. Another possibility is hitting the planet with large comets rich in nitrogen-bearing ammonia: Position the impacts over ice deposits and you might reduce the need for underground nukes considerably. One should be careful, though, not to spray too much of the water released by the impacts straight back into space. That would be wasteful. Mike Carr's ghost would object.

It is clear that true terraforming requires technologies that are almost inconceivable. The almost, though, is important. As McKay's papers, Fogg's textbook, and the Ames workshops show, the almost inconceivable is, in fact, conceivable; we can imagine the terraforming of Mars much more realistically than Julius Caesar could have imagined a jumbo jet, or Socrates a *Saturn V*. That something is conceivable does not, in itself, prove anything. It may simply be that ours is a time in which the technological imagination is far less tightly bound to the state of the technological art than used to be the case. But at the same time, our technology has become capable of operating at planetary scales.

Humankind had never really seen a planet-wide weather phenomenon before *Mariner 9* sent back images of the great dust storm of 1971; within a decade or so, spurred on in part by those observations of Mars, researchers at Ames and elsewhere had shown that a similar atmospheric shroud might be brought about in days by means of a nuclear winter. Though we don't have tools designed for terraforming, we do have many tools of the same general kind, even if they are abacuses to computers. We don't know how to design complex ecosystems, or long-term life-support systems; but we know how to build spacecraft and hydrogen bombs and genetically engineered organisms and greenhouse gas factories. And we know that we can change a planet's atmosphere and climate. No one seriously denies that CFCs have eaten away at the Earth's ozone layer. Though there are, as of this writing, some people who think that anthropogenic increases in carbon dioxide, methane, and CFC levels are not responsible for the global warming seen over the past century, no one thinks that, if we were deliberately to try to warm the Earth with supergreenhouse gases, we wouldn't succeed.

For thousands of years, humanity has increased its power to reshape the Earth. The pharaohs built pyramids; the Dutch built half a country. China and Europe lost their forests; large parts of the American West were irrigated. Today humanity moves more soil around the world than all the rains and rivers combined. At the moment, we are wisely trying to minimize our impact on the Earth; helping the ozone layer to replenish itself, limiting the amount of global warming we are responsible for, attempting to curb our tendency to drive our fellow creatures extinct. In time, though, we may see reasons to take a more active role in things. Sometime in the next few centuries, we can expect to have occasion to move aside an asteroid or comet and save some region of the earth—even, possibly, the Earth as a whole—from a ghastly devastation. Sometime in the next few millennia we will in all likelihood have to choose between adapting to an ice age and averting one.

Civilization—the art of living in environments designed by humans rather than those found in nature—is only about ten thousand years old. Though it is still a fragile thing, and not to be taken for granted, there is no reason to suppose it is near its end. And if it persists to twice its present age, it will be through designing those environments on a planetary scale with both greater care and greater ambition, protecting them from bolts from the blue and changes in the climate and other calamities that would once have been acts of God. If this civilization spreads to Mars, might it not reshape the planet to suit its purposes? And even if Mars is not settled, might it not, in this long run, be reshaped anyway? Robert Zubrin speaks of terraforming Mars as a natural consequence of settlement—making worlds, he argues, is what people do. This is not quite the nineteenth-century assertion that "rain follows the plow"—an assertion that proved horribly untrue in the dust-bowl years—but it's not terribly far off. Others suggest planetary engineering might be undertaken as a grand ecological experiment; Martyn Fogg's book opens with a striking quotation from the physicist Richard Feynman, "What I cannot create, I do not understand." Reengineering might be a moral act, an observant restitution on behalf of the hidden Martian biota, an ecopoiesis for bacteria and plants, rather than a terraforming for humans. It might be a spiritual act, the creation of a new wilderness as solace or penance. Or it might be a work of art: a landscape recrafted the better to be itself and fit the cosmos, a vast Roden Crater.

Terraforming might be economic, scientific, religious, or aesthetic; it might be a mixture of some or all of them; it might be none. The fact that we can imagine the tools that would be necessary does not mean we can know the motivations for using them. The future, like the past, is a foreign country, a place where the connections between people and their values are all too often opaque to us. The various threads that I see tying our world together when I look down the Thames from Greenwich Park were already being woven when Queen Elizabeth I looked out from the same heights: Ships sailed down the Thames and around the world (just), trade prospered across oceans, manufacturing was becoming newly dependent on capital, information was beginning to flow more freely thanks to the printing press. But no one then could have seen where those potentialities would lead: the squat bulk of Reuters, the towers of the banks, the trade routes revealed by the landing lights on aircraft.

Though it flows from the present, the distant future is another world. We cannot know the future for what it is, because as yet it isn't; we cannot know it for what it will be, because by the time it is, we won't be. It is connected to us—its seeds are in us—but we are not connected to it. Though it is human, it is in some ways more alien than a distant but knowable planet can ever be. Mars is tied to us only loosely—you could ignore it for a lifetime and not be thought strange—but it is part of our world, tied to us by Airy and observation, by science and escapism, by novels that use it to articulate today's concerns and paintings that try to inspire our tomorrows. In its small, distant manner it is more a part of our shared world than the undiscovered future downstream can ever be.

The Undiscover'd Country

> "You read us his words, as great as Locar's. You read to us that there is 'nothing new under the sun.' And you mocked his words when you read them—showing us a new thing."
>
> "There has never been a flower on Mars," she said, "but we will learn to grow them."
>
> —Roger Zelazny, "A Rose for Ecclesiastes"

The possibility of life on Mars—whether indigenous or imported, whether past or future—carries with it an obvious corollary. If Mars has life then it must have death. Indeed, death on Mars has a literature all of its own. One of the most famous of Theodore Sturgeon's science fiction stories is "The Man Who Lost the Sea," a vignette of a dying astronaut on the Martian plains. Arthur C. Clarke chose the Sturgeon piece as his contribution to an anthology called *My Favorite Science Fiction Story* and has played a variation on the same theme himself in "Transit of Earth." In a story strongly under the influence of *Scott of the Antarctic,* Clarke's doomed astronaut is Captain Oates in reverse; their spacecraft crippled when the permafrost it landed on gave way, his colleagues have died in order to leave him with enough oxygen to make a long-planned, if fundamentally inconsequential, scientific observation. After this, he too will die.* In Gordon Dickson's novel *The Far Call* and Ludek Pesek's grim children's book *The Earth Is Near,* the protagonists most concerned with Mars are doomed. Robinson's *Red Mars* opens with the murder of the first person to reach the planet.

This is not all that surprising. Mars is a desert, and in America the literature and iconography of the desert are filled with death. Low-

*At around the same time Isamu Noguchi carved a beautiful sculpture called *Planet in Transit,* its black granite redolent of the grave.

ell's Mars was a cosmic reconstruction of this idea, and a grim forewarning of the Earth's inevitable end. Its color, like that of the Painted Desert beyond the forests of Flagstaff, might be "lovely beyond compare; but to the mind's eye, its import is horrible. All deserts, from a safe distance, have something of this charm of tint . . . but this very color, unchanging in its hue, means the extinction of life . . . The drying up of the planet is certain to proceed until its surface can support no life at all." That deserts are also places that live is a wonder precisely because of the hardship involved. Life's presence does not alter the essential deathliness of the place; it highlights it, whether it's a mouse burrowing in the roots of a dwarf juniper or a set of canals that girdles a planet. To Lowell the canals were a great triumph of engineering, a testament to a superior and peaceful civilization. But they were also futile. Later writers, from Ray Bradbury to Philip K. Dick to J. G. Ballard, would echo the theme; Mars was a place where entropy was on the rise.

It was, in part, a sense of the fragility of life in the desert that led NASA to ensure that the *Viking* landers (perhaps the least living things mankind has ever made) were assembled in sterile conditions, purged with hot helium, and baked in vast ovens for days. Though this had an immediate practical purpose—if the landers were not sterile their life-detection experiments might end up discovering common-or-garden earthly bacteria that had simply happened to come along for the ride—it was also driven by the ethical concern that a few earthly bacteria might wreak havoc in a marginal Martian ecosystem. It was, in the words of the Caltech exobiologist Norman Horowitz, one of the people working on the life detection package, "a monument to a Mars that never existed." The deadest thing human hands have ever built became something very like a tombstone on the planet's surface.

If deserts are deathly, so, too, is science fiction in general. Genre expectations within the literature of the fantastic divide the worlds it describes into categories loosely analogous to the two "other worlds" of everyday life. The worlds of fantasy are arbitrary, but with clear significance; they correspond to the other world of our dreams. The worlds of science fiction, on the other hand, are frequently portrayed as necessary, beyond our current boundaries in space or time but related to our world by an imagined history. They correspond to the other world we face by necessity: the world to come. A decade ago, ruminating on the fact that so much then-current science fiction made this peculiarly explicit—there was a vogue at the time for stories that started after the end of, if not the world, at least the earth—the critic John Clute wrote, "In our hearts, most of us in 1992—writers and readers alike—read sf in the secret conviction that the genre is a body of fairytales about the afterlife."

If our natural ways of thinking about other worlds are as dreams or afterlives, then Mars is clearly the second. Mars no longer has any of the arbitrariness of the dream about it. We know it far too well. It has the same things in it that our everyday physical world has—sand, sky, sunsets. The one thing that is missing is us. Freud claimed, influentially but unconvincingly, that people cannot truly believe in their own death; Goethe said much the same. Imagining one's death, according to this logic, meant imagining one's survival as a witness to that death, and this meant one was not really imagining a world in which one was absent. In fact, though, one can imagine a world in which one is dead: One simply cannot grace the imagining with full subjectivity. In the way we are forced to imagine them, death and Mars are much alike.

Through naming its features after dead astronomers and writers, the IAU has made Mars yet more like a cemetery. So has the American custom of renaming their meticulously sterilized spacecraft in honor of dead colleagues after they land on Mars. And this association with death, far from making Mars more distant, more unapproachable, makes it closer and more real. The Chinese anthropologist Fei Xiaotong, visiting America in the 1940s, was sensitive to a lack of the ties that bound together past, present, and future in his childhood China, a lack he summed up by describing America as "a world without ghosts." With our memorials we are giving dead Mars its first ghosts. And that binds us to it in new ways. I have felt close to Mars in many ways while writing this book. I have looked at meteorites that came from Mars and lost myself in the writings of people who have imagined it in greater detail than I would have thought possible. I have relaxed in Tom Meyer's backyard a few miles out of Boulder, a happy guest in the heart of the Mars Underground as we lie on our backs and gaze up at the evening sky. I have pored over maps and felt that I could walk out into them. I've watched Bob Zubrin make people really—really—believe they were on their way there. I have looked out of aircraft windows and almost felt I could imagine what the bulk of Olympus Mons would look like off our wing, filling up a state's worth of the American West and shouldering aside the sky; I have watched the planet itself from a parking lot in Flagstaff. I have listened to my godson Jack talking in his five-year-old matter-of-factness about what he or I might wish to do there. I have looked out across the beautiful curve of the Gulf of Lyons from the foothills of the Pyrenees and seen it as the shoreline of a flooded basin on a terraformed paradise (and I have got out my compasses, calculated the radius of curvature and tried to work out which basin might best fit the bill: something a bit bigger than Gusev would do nicely).

But I don't think I have ever felt closer than in the lobby of the Smithsonian Air and Space Museum in Washington, D.C. On the floor is a *Viking* lander, a flight spare almost identical to the two that sit on Chryse and Utopia; on its side is a plaque that identifies it as the Thomas Mutch Memorial Station, named in honor of the geologist in charge of the lander science program, who died in 1980. Displayed with the plaque is the explanation that while the lander is the property of the museum, the plaque is merely on loan; one day NASA will take it away and mount in on the lander's dusty twin, a few hundred miles from the mouth of Kasei Vallis. There is no date; there is no policy; there is no guarantee. But the things we say on gravestones have a special value, a requirement for commitment. You're not meant to touch the plaque. I did. It meant more about Mars than a meteorite.

John Clute made his comments about science fiction as a set of fairy tales about the afterlife in a review of Stan Robinson's *Red Mars,* and it specifically excepted Robinson. Though the tale may be set after many of us are dead (his characters arrive at Mars around Robinson's seventy-fifth birthday), the book and its sequels are committed to the idea that the history it relates is, for good or ill, continuous with that which we inhabit—the same commitment that the Mutch plaque makes. The only worlds that end between now and then are the worlds that end with every death; the political and economic constraints of today are carried forward to tomorrow and outward to Mars. The new planet is not the New World that some of its would-be settlers want it to be; it is a new part of our Old World.

I sat next to Stan Robinson at the NASA Ames terraforming meeting in the autumn of 2000. At the podium, Charles Cockell was talking about his biological research at Haughton crater in Canada. The most interesting environments he had found were what he called micro-oases. Four feet across, or a bit more, and more than ten inches deep (the permafrost comes close to the surface), the micro-oases are tiny islands of rich plant life in an environment where 99.7 percent of the surface is bare rock (the figure is precise: Charles measured it himself). These oases, he told us, are the sites where large animals have died. Sometimes a musk ox will wander into Haughton and fail to wander out. Very slowly, its body will decay and enrich the soil. Wind-borne seeds will sprout in it; lemmings will find the soil there softer than its surroundings and chose it for their burrows. Once the lemmings are established, the oasis takes on a life of its own, their droppings replenishing the original bequest of nutrients. How long these oases last is not yet known. Decades, certainly; centuries, possibly. Robinson and I looked at each other; Nathalie Cabrol, a few seats away, had the same idea; so did a number of others across the auditorium—you could see them lifting their heads, or scribbling a note, or nudging their neighbors. Here, we all thought, was a way to put the poetry into ecopoiesis.

Take away the water and a human body contains some forty pounds of interesting nutrients. On the Earth, we take these for granted. On Mars, they would be worth something. A human body, carefully desiccated, might in a few decades, or a century, be sent to Mars for the equivalent of a few thousand dollars; dead people are far cheaper to send there than live ones. Entry packages could be designed that got the remains down to the surface in one piece, rather than spreading them through the atmosphere, and implanted them a little way into the regolith, so the wind would not carry them away. A little microchip would carry the deceased's name and some version of his or her life story; a carefully tended database would record where each of the loved ones had ended up. And that would be it, until the ecopoiesis began. Then, when warmth and water returned to the surface, life would begin to spread and would start to make use of resources. The seeds of the future would find their nutrients and would make use of them.*

Would people really wish to be buried on Mars? It is far from home, by any standards. But some might make the choice. When trying to find figures on mineral extraction for the discussion of supergreenhouse gases in the previous chapter, I came across a brief biography of Frank Taber. Taber was a man of the West, born in Oregon, died in Montana. He married his college sweetheart and joined the Bureau of Mines before moving to Bitterroot Valley and running what was then the largest fluorspar mine in the world (which is why the search engine offered him up to me). He flew his own plane and relished the new insights he got into the land from on high; but he was also happy scratching a rock with his pocket knife and examining the results with a hand lens. He gazed at the moon and stars as Gene Shoemaker had; I imagine he would have enjoyed a glass of bourbon with Dave Scott. I'd have liked to have met him.

I mention this because Taber's biography is a virtual tombstone on a site run by the Celestis Foundation. Celestis offers its clients burial in space. A sample of Taber's remains, along with those of twenty-five others, made up the second Celestis payload, launched in 1998 and now in orbit around the Earth. The Web site will even tell you where to look to see the satellite. Gene Shoemaker went further, being buried, in part, on the moon; some of his ashes were on board the Lunar Prospector spacecraft when it was crashed near the moon's south pole (deliberately) in June 1999. The spacecraft bore a small inscription from *Romeo and Juliet:*

*In his magisterial history *Landscape and Memory,* Simon Schama devotes a short chapter to the aspect of humanity that Thomas Cole and others were most willing to work into their wildernesses: the sign of the cross, frequently in the form of a headstone. It's an aspect of their art that has not been picked up much in portrayals of Mars; though Bonestell painted a Martian burial, there is no cross in evidence, just a flag. The title of Schama's chapter is "Vegetable Resurrections."

> And when he shall die,
> Take him and cut him out in little stars,
> And he will make the face of heaven so fine
> That all the world will be in love with night,
> And pay no worship to the garish sun.

For Gene, the moon was the right choice. Mr. Taber, though, might have chosen Mars if the option had been available. So might many more in years to come, if given the chance. They would be little capsules of earth—of Earth's earth—moved to Mars. Their locations would be mapped and there they would sit, awaiting the dream of a new life. To some ears this may sound macabre, but the funeral practices of other times and places often do. It might sound like a desecration, the dumping of human waste on a pristine land.* But this is much more like a sacralization than a desecration. The transformation of body to micro-oasis would be an ecopoietic sacrament.

Terraforming is a mighty dream, but it can be started in small ways. Chris McKay has recently been narrowing his focus down to the first vital step: putting the first living thing from Earth onto Mars. Not an astronaut. A flower. Back in the 1970s Penny Boston showed the Underground that you could grow things in pretty Martian conditions; the little glass Mars jars with thin carbon-dioxide atmospheres and even thinner radishes were one of the main attractions in the attic room in the physics building that they called their own. But there is a difference between knowing something and seeing it done. *Sojourner*'s makers knew their little rover could roll around a rocky sandpit—they'd seen it do so on a back lot at JPL. But it made all the difference to them and the world to see it done on Mars. The same applies to the flower. *Sojourner* was, in NASA jargon, a technology demonstrator. The flower, says McKay, would be a biology demonstrator. He imagines a tiny greenhouse on a lander; it might use Martian soil, chemically neutralized, or it might

*On the subject of waste, it must be admitted that in the Arctic a latrine site will do almost as well as a dead musk ox; this is why the punctilious researchers at Haughton ship all their excreta out at the end of each field season.

not; it might extract some water from the Martian atmosphere or from ground frost, or it might not. But it would use Martian air and Martian sunlight, and from those a plant would grow.

At present, NASA still has a policy of cleaning its Mars-bound spacecraft almost as thoroughly as it cleaned the *Vikings* in order to protect the environment. This makes a lot of sense in many ways, especially for spacecraft looking for life. But at sometime this planetary protection will have to be relaxed. After all, the rhetoric of Mars exploration, as spoken by space program administrators and politicians, is that one day people will go there—the question is always said to be one of when, not if. Keeping a human presence utterly sealed off from the environment would be completely impossible. So at some point, we have to let life loose on Mars. The plant would be the first step.

It wouldn't be a rose. Roses are too slow to grow, too hard to tend, too woody and whatever else. But in my mind's eye I see it as a rose, a rose that reminds me of the rose garden in Greenwich Park, of the boutonniere I wore on my wedding day, of the rose bed my father dug for my mother in the garden of the last house he ever lived in, of the roses my wife Nancy looks after on the back step of our gardenless apartment. I see its bud opening and the faint sunlight on its petals; I imagine water evaporating from its leaves, and more being drawn up from the gently irrigated soil, moistened by canals a trillionth of the size of Lowell's planetary monstrosities. I see hundreds or thousands of similar flowers carefully tended in earthly classrooms by children following the Mars flower's progress, their growth minutely measured as life unfolds in parallel on two different planets. I see it fading in a strange way, withering without rotting in a place where there are no bacterial worms to eat away at it, where a rose cannot be sick—but must die, nonetheless, as the seasons demand. Planetary protection be damned, I see it dropped from the greenhouse to the soil below. There the UV and peroxides will work their dark magic; the rose will be distilled into a few grams of carbon dioxide, water, and a little nitrogen.

Those molecules would spread across the planet. On Earth the atoms are used to life; life takes them in and spits them out repeatedly. Your every breath contains atoms breathed in and out by your most distant ancestors; each new tree takes bequests from the world's crematoria into itself. The great dance cycles on as it has for billions of years; but now it can be started again. When people go to Mars, their breath will carry faint traces of your own, because your breath is spread around the world. When plants grow on Mars they will take up that breath and fix it in new soil. When people die on Mars, they will leave there atoms that once made up a tiny part of your ancestors. Even if no terraforming takes place, a transfer will still have occurred.

And if the greater changes come, then, by and by, the faint traces of that first rose and the exhalations of all humanity will pass through the tissues of a living world. The carefully charted constellation of micro-oases will come to life like a vegetable empire as seeds that fall on the grave sites blossom and grow. Perhaps no one will come by to read the microchip epitaphs for centuries, if at all; perhaps the life of the planet will be reserved for plants and microbes, avoiding the vast trouble of releasing the levels of oxygen needed by animals. Either way, the oases will still grow, even if the flowers are unseen, the fruit unpicked, the herbs unsmelled: saxifrage in cracked stone, sweet alpine strawberries, rosemary for remembrance.

Dead as we may be, breath by breath we will go to Mars.

Acknowledgments

My thanks go first and foremost to the people who spared the time to talk to me about their involvement with Mars at various times during the gestation and creation of this book. So thank you to Jim Cutts, Steve Squyres, Steve Saunders, Norm Haynes, Charlie Cockell, Bob Haberle, Don Scott, Martyn Fogg, Carol Stoker, Jim Zimbelman, Ken Edgett, Edmond Grin, Jeff Moore, Julian Hiscox, Kim Poor, Matt Golombek, Alfred McEwen, Carl Pilcher, Peter Schultz, Herb Frey, Jeff Kargel, Mary Chapman, Larry Soderblom, Norm Sleep, Dave Gibbons, Mike Carroll, Phil Christensen, Kevin Zahnle, George McGill, Tracey Clegg, Jim Rice, Bruce Banerdt, Ken Tanaka, Virginia Gulick, Bruce Jakosky, Richard Poss, Baerbel Lucchitta, Larry Lemke, Bill Clancey, Michael Sims, Ben Clark, Dave Morrison, Steve Clifford, Ray Batson, Pascal Lee, Carlos Miralles, and Bob Zubrin.

Among the people particularly generous with various mixtures of time, insight, and help were Chris McKay, Penny Boston, Bill Hartmann, Baerbel Lucchitta, Mike Carr, Tim Parker, Ken Edgett, Peter Smith, Nathalie Cabrol, Geoff Briggs, Vic Baker, Randy Kirk, David Smith, and Don Wilhelms, who provided an extremely helpful archive of papers relating to *Mariner 9;* they also kindly read and commented on drafts of some parts of the book. Bruce Murray did a great deal to get me interested in planetary science in the first place and also kindly explained to me why this wasn't the book to write.

When I started writing this book, neither Dave Scott nor Mert Davies met the IAU criteria for having craters named after them on Mars; around the time I finished it they both, sadly, became eligible. I'm grateful not only for their help, but also for having had the chance to meet them.

Jay Inge, Pat Bridges, and their colleagues deserve special thanks for the maps that inspired this approach to writing about Mars. Stan Robinson did a great deal to make Mars feel real enough to write about, as well as being generous with time and encouragement once I'd gotten started.

Thanks are also due to Ruth Banger at the Department of Earth Sciences in Cambridge, the friendly and very efficient staff of Cambridge's Scientific Periodicals Library, Peter Hingley at the Royal Astronomical Society, Andy Sawyer at the Science Fiction Foundation collection in Liverpool, and the staffs at the British Library and Cambridge University Library. Adrienne Wasserman at the USGS Flagstaff was an enormous help in supplying copies of maps. Jennifer Blue in Flagstaff was an enthusiastic source on the IAU nomenclature process that she administers. Anne Cairns and her colleagues at the Geological Society of America were very helpful, as were their counterparts at the thirtieth Lunar and Planetary Science Conference. Among many people who helped in NASA press offices particular thanks go to Kathleen Burton, Mary Hardin, Cynthia O'Carroll, and Doug Isbell.

The following permissions for quotation are gratefully acknowledged: from William Fox courtesy of the author; from *Seinfeld* courtesy of Castle Rock Entertainment © 1993, all rights reserved; from John McPhee courtesy of the author; from Ray Bradbury courtesy of the author and Abner Stein; from Frederick Pohl's recollection of Isaac Asimov courtesy of *Locus* magazine; from John Clute courtesy of the author; from Robert Harbison courtesy of MIT press; from Frederick Turner courtesy of Saybrook Publishing; from Scott McCloud courtesy of the author; from Robert W. Service courtesy of the author's estate, administered by William Karsilovsky; from W. H. Auden courtesy of Faber and Faber and Random House; from Andrew Brown courtesy of the author; from Frank Herbert courtesy of the Herbert Limited Partnership; from

Buffy the Vampire Slayer courtesy of Twentieth Century Fox, from Sir Arthur C. Clarke courtesy of the author; from *The Matrix* courtesy of Warner Brothers; from Benton Clark courtesy of the author and the American Astronautical Society, which originally published the article as paper AAS 78-156 in "The Future United States Space Program," Johnston, Naumann, and Fulcher, Eds., © 1979; from "Life on Mars," words and music by David Bowie © 1971, permission courtesy of EMI Music Publishing Ltd/Moth Music/Tintoretto Music/Chrysalis Music Ltd/RZO Music Ltd; from Walter M. Miller Jr. courtesy of Random House; from Roger Zelazny courtesy of The Pimlico Agency Inc., agents for the Estate of Roger Zelazny; from David Hockney courtesy of Rogers, Coleridge, and White.

My thanks also to the copyright holders and suppliers of the various illustrations. The endpapers, which combine *Viking* and MOLA data, are © Ralph Aeschilman 2002. Figures 1 and 2 courtesy of the Royal Astronomical Society; figure 3 courtesy of NASA/JPL/Caltech; figures 4–7 courtesy of NASA/JPL/Caltech and the Regional Planetary Information Facility at University College London; figures 8 and 10 courtesy of the Lowell Observatory; figures 9, 11, and 12 courtesy of the Astrogeology Team, USGS, Flagstaff, Arizona; figure 13 courtesy of NASA/JPL/Caltech; figure 14 courtesy of NASA/JPL/GSFC/MOLA team; figures 15–18 courtesy of NASA/JPL/MSSS; figures 19 and 20 courtesy of the copyright holder, Bonestell Space Art; figure 21 courtesy of William K. Hartmann; figure 22 courtesy of Founders Society Purchase, Robert H. Tannahill Foundation Fund, Gibbs-Williams Fund, Dexter M. Ferry Jr. Fund, Merrill Fund, Beatrice W. Rogers Fund, and Richard A. Manoogian Fund—photograph © 1985 The Detroit Institute of Arts; figure 23 courtesy of NASA and the artist; figure 24 provided by the Department of Earth Sciences, Cambridge University; figure 25 courtesy of NASA/JPL/Caltech/Univeristy of Arizona; figure 26 courtesy of U.S. National Archives; figure 27 courtesy of NASA, digital image © 1999, Michael Light; figure 28 courtesy of Michael Carroll, montage by Julian Humphries. My thanks to David Cutts and Richard Riddick at Othens for photographs and scans.

At the Wylie Agency Sarah Chalfant was the most excellent of agents, and Helen Allen and Sonesh Chainani were endlessly helpful. Joshua Kendall at Picador and Clive Priddle at Fourth Estate were supportive and insightful editors; Kate Balmforth, Julian Humphries and Mitzi Angel performed the wonders that turn an edited text into an actual book. Ilsa Yardley was an excellent copy editor.

Editors who have generously tolerated or indeed encouraged ruminations on matters Martian over the years (not to mention subsidizing them) include Matt Ridley, Anthony Gottlieb, John Peet, Bill Emmott, Sarah Richardson, Fred Guterl, Michael Elliott, David Kuhn, Rick Hertzberg, Tony Reichhardt, Jeff Mervis, Richard Stone, Gabrielle Walker, Justin Mullins, and Jessie Scanlon. There was also moral, intellectual, and practical support at various times and places from fellow writers and others: David Chandler, Jim Oberg, Richard Wagner, Keith Cowing, Charles Seife, Faye Flam, Leonard David, and Barbara Sprungman; Tom Standage and Roger Highfield (and Roger Highfield's neighbor); Mark Roberts and Geoff Carr; Edmund Fawcett and Jim Flint, who provided a couple of off-the-cuff insights; Bill Fox, who seemed to have been almost everywhere first; Simon Schaffer, without whose influence everything would be rather different; John Heilemann; Adrian Johns and Alison Winter; Ian Watson, Geoff Landis, and Greg Benford; Dave Gibbons; James and Sandy Lovelock; Matt, Julie, Jack, Kate, Anne, and Graham Bacon; and my extended family on two continents, with a special mention for Bob's Buffalo Clan, who brought me back to Earth very effectively within twenty-four hours.

One early passage stems directly from a stroll taken with Kim Miller, who is one of the many reasons I must save the greatest part of my gratitude for Nancy Hynes, my most supportive companion throughout the writing, a source of challenging insights and the possessor of a great eye for art, a wonderful reader, and the provider of the rose now sitting on my desk.

References, Notes, and Further Reading

The following is intended as a general guide to where the claims in the text come from and thus to where the eager reader can learn more; it's broken down by chapter, with references to the bibliography provided in the form of "Author (date)."

Introduction

The maps in question are Hall, Davis, and Inge (1985).

Part 1

Greenwich Inasmuch as one afternoon is described here, it is that of Friday 3 March 2000. The material on Airy comes from his autobiography and his entry in the *Dictionary of Scientific Biography*, supplemented by chats with Simon Schaffer and my neighbor Tom Standage (whose recent book *The Neptune File* deals with other aspects of Airy's career). Information on the Washington meridian conference comes from Harrison (1994) and Jennings (1999).

A Point of Warlike Light The history of Mars as seen from Earth follows Sheehan (1996); some complementary material is from Moore (1998). A very good account of Martian observations in the wider context of planetary astronomy and the search for life is to be found in Dick (1996). In matters of classically derived nomenclature the definitive source is Blünck (1977) (also the source of the quote from Schiaparelli). Jennifer Blue of the USGS provides an on-line gazetteer for place names on Mars and across the solar system at http://wwwflag.wr.usgs.gov/nomen-bin/search.cgi. The Tennyson is from "Locksley Hall." The Lowell quotes come from Lowell (1896).

Mert Davies's Net This relies almost entirely on a day spent with Davies in December 1999, and on numerous papers and other documents he kindly gave me. Conversations with Bruce Murray were also helpful; a sense of the excitement of their early work together is nicely caught in *The View from Space*, Davies and Murray (1971).

The Polar Lander Details of the mission are taken from the NASA "*Mars Polar Lander/Deep Space 2* Press Kit," which is reprinted, along with other relevant material, in Godwin (2000); the analysis of the mishap comes from the Casani report, JPL document D-18709.

Mariner 9 Technical details on the spacecraft are from Ezell and Ezell (1984). Don Wilhelms provided a treasure trove of relevant USGS documents, including floor plans, work rosters, and progress reports for the *Mariner 9* television team. The Caltech symposium formed the basis of Bradbury et al. (1973), and Bruce Murray tells the story in his memoirs, *Murray* (1989). The wonder of seeing Mars anew is caught in Hartmann and Raper (1974). I'm grateful for conversations on *Mariner 9* with Mike Carr, Jim Cutts, Mert Davies, Bruce Murray, Stephen Saunders, Larry Soderblom, Don Wilhelms, and Jurrie van der Woude.

The Art of Drawing The all-important key reference here is Batson, Bridges, and Inge (1979)—the 1:5,000,000 atlas of Mars. Lunar cartography and the roots of the airbrush method are covered in Wilhelms (1993) and, in greater detail, in Kopal and Carder (1974); for this and other aspects of planetary mapping see Greeley and Batson (1990). This chapter was largely based on conversations with Jay Inge, and also with Ray Aeschliman, Pat Bridges, and Ray Batson.

The Laser Altimeter The shuttle-linked decline of planetary exploration from the 1970s is chronicled in Burrows (1990) and in Burrows (1998). Material on MOLA comes from conversations with David Smith; Geoff Briggs helped with perspective, as did Bruce Murray.

Part 2

Meteor Crater A helpful product from the gift store is Smith (1996); Mark (1995) provides more detail and context. Herbert Frey showed me the sliding color approach to discovering rimless craters in the MOLA data set.

"A Little Daft on the Subject of the Moon" For Gilbert, see Pyne (1980); Goetzmann (1966) is useful on the general history of the scientific exploration of the West. The quotation from Dutton is from Dutton (1882). Gilbert's paper is Gilbert (1896). The material on Shoemaker comes from Wilhelms (1993), Levy (2000), and Shoemaker (1999).

An Antique Land Nomenclature is discussed in Blünck (1974) and on-line at the USGS Gazetteer (see *A Point of Warlike Light* above). Descriptions of the Martian surface follow Carr (1982) and Kieffer et al. (1992). Bradbury quote from Bradbury et al. (1973); Turner quote from Turner (1988). The anecdote about the pinkening of the sky is from Poundstone (1999), as is most of this book's Sagan anecdotage.

Maps and the Multiple Hypotheses The best general account of planetary geological mapping is Don Wilhelms's chapter in Greeley and Batson (1990). The mid-1980s geological maps of Mars are Scott and Tanaka (1986), Greely and Guest (1987), and Tanaka and Scott (1987). There are small reproductions in Kieffer et al. (1992). McCloud quote from McCloud (1994). This chapter draws a great deal on conversations with Dave Scott, Don Wilhelms, and Kenneth Tanaka, and on material covered in a number of academic papers by Wilhelms and Squyres, Sleep, Frey, and others, details of which were clarified by discussions with the authors.

The Artist's Eye For the art itself see Hartmann and Miller (1991) and Hartmann et al. (1984). The discussion of the art draws on a number of conversations with Richard Poss; it also benefits from Schama (1996), Hughes (1997), and Andrews (1999). Thanks to Dave Gibbons and Alan Moore for the quote from the unpublished script for the Martian sequence of *Watchmen* (Moore and Gibbons (1986)). Crater dating is very helpfully discussed in Carr (1982); a number of talks with Bill Hartmann were invaluable.

Layers Based largely on conversations with Edgett, Schultz, and Lucchitta, and associated technical papers. Background on the taciturn Dr. Malin is from Cooper (2000). The Hockney quotation is from Hockney and Joyce (1999).

Part 3

Malham The geology of Yorkshire draws on Fortey (1993) and on a visit to the visitors' center at Malham. The great climate paper is Pollack (1979). Oreskes (1999) influenced the view of geology throughout this book.

Mike Carr's Mars The main strands of Carr's thought, developed in a series of technical papers, are brought together in Carr (1996). This chapter follows the original papers, as clarified in talks with Carr.

Reflections The *Martian Inca* anecdote comes from a conversation with Watson. The ideas about Mars in science fiction are influenced by Baxter (1996b), by Dick (1996), by Westfahl (2000), and by Brian Stableford's essay on Mars in Clute and Nicholls (1993).

Shorelines Conversations with Parker, Baker, and Kargel shaped this chapter. Scablands material comes from Baker (1982) and Gould (1980). The episodic-ocean paper is Baker et al. (1991).

The Ocean Below This reflects the experience of the 2000 Lunar and Planetary Science conference, and conversations there and elsewhere with Head, Parker, Baker, Clifford, and others.

"Common Sense and Uncommon Subtlety" The sources are conversations, especially with Cabrol and Haberle, and attendant technical papers; the Malin/Edgett gullies press conference was a helpful Web cast. The use of the quote from Lowell (1906) as a title was inspired by Kevin Zahnle.

Part 4

Buffalo The Buffalo material comes from the meeting itself. The account of landing-site selection for *Viking* draws on Ezell and Ezell (1982) and Poundstone (1999); that for *Pathfinder* draws on Raeburn (1998) and on technical papers by Golombek et al.

Putting Together a Place Discussions with Peter Smith shaped this chapter, as did looking at the work of Michael Light and David Hockney. The documentation of *Pathfinder*'s placehood is in *Science* 278, No. 5344, 5 December 1997.

The Underground Material on the Mars Underground comes from talks with Chris McKay, Boston and Tom Meyer, at the Case for Mars IV, V, and VI, and elsewhere, as does material on McKay's and Boston's scientific work. The account of ALH 84001 draws on those in Taylor (1999) and Achenbach (1999).

Bob Zubrin's Frontier The key source here is Zubrin (with Wagner) (1996), supplemented by reporting at the first and second conventions of the Mars Society.

Mapping Martians The Haughton Mars Project has an excellent Web site, http://www.arctic-mars.org. Bill Clancey's papers are at http://home.attnett/WJClancey. Further information from talks with Lee, Clancey, Zubrin, and Cockell.

Part 5

Symbols of the Future This is really my own interpretation of the science fiction discussed in the chapter, mixed with a critical reading of Zubrin's arguments in Zubrin (with Wagner) (1996) and elsewhere, and discussions of exploration with Charles Cockell. Schivelbusch (1987) provided an understanding of railways.

Gaia's Neighbor Jim Lovelock's part of the story is in his memoir, Lovelock (2000). Oberg (1981) is a groundbreaking book on the subject of terraforming, and Fogg (1995) is still pretty much authoritative, though some progress (in supergreenhouse gases, most notably) has been made since. Achenbach (1999) captures the great terraforming debate at the Mars Society's founding convention.

The Undiscover'd Country Patricia Nelson Limerick discussed responses to American deserts in Limerick (1985) and introduced me to Fei Xiaotong's "World without Ghosts" in Arkush and Lee (1993). The review of *Red Mars* from which the Clute quotation is taken can be found in Clute (1995). Frank Taber's biography is on the Celestis Web site, http://www.celestis.com.

Bibliography

This contains the books and articles referred to directly in the text and in the notes and references; a fuller bibliography containing more technical papers is available on the Web.

Achenbach, Joel (1999) *Captured by Aliens: The Search for Life and Truth in a Very Large Universe,* Simon and Schuster—both perceptive and funny, often simultaneously

Airy, George Bidell (Wilfrid Airy, ed.) (1896) *The Autobiography of Sir George Biddell Airy,* Cambridge University Press

Aldiss, Brian W. (1986) "The Difficulties Involved in Photographing Nix Olympia," in *Best SF Stories,* Gollancz, 1988

Aldiss, Brian W. and Roger Penrose (1999) *White Mars or, the Mind Set Free: A 21st-Century Utopia,* Little Brown

Alpers, Svetlana (1983) *The Art of Describing: Dutch Art in the Seventeenth Century,* University of Chicago Press

Andrews, Malcolm (1999) *Landscape and Western Art,* Oxford University Press

Arkush, R. David and Leo L. Lee (1993) *Land Without Ghosts: Chinese Impressions of America from the Mid-Nineteenth Century to the Present,* University of California Press

Baker, Victor (1982) *The Channels of Mars,* Adam Hilger, Bristol

Baker, V. R., R. G. Strom, V. C. Gulick, J. S. Kargel, G. Komatsu, and V. S. Kale (1991) "Ancient Oceans, Ice Sheets, and the Hydrological Cycle on Mars," *Nature,* 352, 589–94

Batson, R. M., P. M. Bridges, and J. L. Inge (1979) *Atlas of Mars: The 1:5,000,000 Map Series,* NASA—a world between covers

Baxter, Stephen (1996) *Voyage,* HarperCollins—a novel in which Spiro Agnew's plans for Mars come true

Baxter, Stephen (1996) "Martian Chronicles: Narratives of Mars in Science and Science Fiction," *Foundation* 68, Autumn 1996

Bear, Greg (1993) *Moving Mars,* Legend, London

Benford, Gregory (1999) *The Martian Race,* Aspect/Warner—an expansion of a novella written with Elisabeth Malartre

Benjamin, Walter (1929) "The Work of Art in the Age of Mechanical Reproduction," in *Illuminations,* ed. Arendt, 1968, Schocken, New York

Bisson, Terry (1990) *Voyage to the Red Planet,* William Morrow

Blünck, Jurgen (1977) *Mars and Its Satellites: A Detailed Commentary on the Nomenclature,* Exposition Press, Hicksville, New York

Bova, Ben (1992) *Mars,* Bantam

Bradbury, Ray (1951) *The Martian Chronicles,* Doubleday—a crucial fiction

Bradbury, Ray et al. (1973) *Mars and the Mind of Man,* Harper and Row

Brown, Andrew (2001) "Under the Surface" in *Granta 74*

Burroughs, Edgar Rice (1912) *A Princess of Mars* (Ballantine edition, 1963)—if this is to your taste see also *The Gods of Mars, Warlord of Mars,* etc.

Burrows, William E. (1990) *Exploring Space: Voyages in the Solar System and Beyond,* Random House

Burrows, William E. (1998) *This New Ocean: The Story of the First Space Age,* Random House

Byerly, Radford (ed.) (1992) *Space Policy Alternatives,* Westview, Boulder—features an excellent essay by Patricia Nelson Limerick

Carr, Michael (1981) *The Surface of Mars,* Yale University Press

Carr, Michael (1996) *Water on Mars,* Oxford University Press

Clark, Benton C. (1978) "The *Viking* Results—The Case for Man on Mars," in The Future United States Space Program, *38* (AAS 78–156)

Clarke, Arthur C. (1952) *The Sands of Mars,* Harcourt Brace, New York

Clute, John (1995) *Look at the Evidence: Essays and Reviews,* Liverpool University Press

Clute, John and Peter Nichols (eds.) (1993) *The Encyclopedia of Science Fiction,* Orbit

Davies, Merton E. and Bruce C.Murray (1971) *The View from Space,* Columbia University Press

Davis, William Morris (1926) "The Value of Outrageous Geological Hypotheses," *Science, 63* 463–8

Dick, Philip K. (1964) *Martian Time Slip,* Ballantine

Dick, Steven J. (1996) *The Biological Universe: The Twentieth-Century Extraterrestrial Life Debate and the Limits of Science,* Cambridge University Press—an all-around excellent guide

Dutton, Clarence (1882) *Tertiary History of the Grand Canyon District,* USGS (separate Atlas published alongside features, wonderful artwork by Holmes)

Ezell, Edward Clinton and Linda Neumann Ezell (1984) *On Mars: Exploration of the Red Planet 1958–1978,* NASA

Fogg, Martyn J. (1995) *Terraforming: Engineering Planetary Environments,* Society of Automotive Engineers

Fortey, Richard (1993) *The Hidden Landscape: A Journey into the Geological Past,* Jonathan Cape

Fox, William L. (2001) *View Finder: Mark Klett, Photography and the Reinvention of Landscape,* University of New Mexico Press

Gilbert, Grove Karl (1896) "The Origin of Hypotheses, Illustrated by the Discussion of a Topographic Problem," *Science,* III, 53, 1–13

Godwin, Robert (ed.) (2000) *Mars: The NASA Mission Reports,* Apogee Books, Burlington, Ontario—a compendium of NASA press kits and postlaunch reports

Goetzmann, William H. (1996) *Exploration and Empire: The Explorer and the Scientist in the Winning of the American West,* Knopf

Gould, Stephen Jay (1980) *The Panda's Thumb,* W. W. Norton

Greeley, Ronald and J. E. Guest (1987) Geologic Map of the Eastern Equatorial Region of Mars, USGS (I-1802-B)

Greeley, Ronald and Raymond M. Batson (1990) *Planetary Mapping*, Cambridge University Press

Hall, Barbara J., Susan L. Davis, and Jay L. Inge (1985) Shaded Relief Map of Mars, USGS (I-1618)

Harbison, Robert (1977) *Eccentric Spaces*, Knopf

Harrison, H. M. (1994) *Voyager in Time and Space: The Life of John Crouch Adams*, The Book Guild, Lewes, Sussex

Hartmann, William K. and Odell Raper (1974) *The New Mars: The Discoveries of Mariner 9*, NASA

Hartmann, William K. and Ron Miller (1981) *The Traveler's Guide to the Solar System*, Macmillan, London; revised and expanded as *The Grand Tour: A Traveler's Guide to the Solar System*, 1991, Workman Publishing, New York

Hartmann, William K., Ron Miller, and Pamela Lee (1984) *Out of the Cradle: Exploring the Frontiers Beyond Earth*, Workman Publishing, New York

Heinlein, Robert (1949) *Red Planet*, Charles Scribner's Sons

Heinlein, Robert (1961) *Stranger in a Strange Land*, Putnam's; restored edition 1991

Hockney, David and Paul Joyce (1999) *Hockney on "Art": Conversations with Paul Joyce*, Little, Brown

Hoyt, William Graves (1976) *Lowell and Mars*, University of Arizona Press

Hughes, Robert (1997) *American Visions: The Epic History of Art in America*, The Harvill Press

Jennings, Charles (1999) *Greenwich: The Place Where Days Begin and End*, Abacus

Kieffer, Hugh H., Bruce M. Jakosky, Conway W. Snyder, and Mildred S. Matthews (eds.) (1992) *Mars*, University of Arizona Press—114 authors provide the definitive post-*Viking* pre-*MGS* view of Mars

Kopal, Zdene and Robert W. Carder (1974) *Mapping of the Moon: Past and Present*, D. Reidel, Dordrecht

Landis, Geoffrey A. (2000) *Mars Crossing*, Tor Books—a Martian adventure in the manner of Shackleton

Levy, David (2000) *Shoemaker by Levy: The Man Who Made an Impact*, Princeton University Press

Lewis, John S. (1996) *Rain of Iron and Ice,* Addison-Wesley

Light, Michael (1999) *Full Moon,* Jonathan Cape (with the Hayward Gallery)—the most beautiful look yet at another planetary surface

Limerick, Patricia Nelson (1985) *Desert Passages: Encounters with the American Deserts,* University of New Mexico Press

Lovelock, James (1979) *Gaia: A New Look at Life on Earth,* Oxford University Press

Lovelock, James (2000) *Homage to Gaia,* Oxford University Press

Lowell, Percival (1896) *Mars,* Longmans, Green and Co., London and Bombay

Lowell, Percival (1906) *Mars and Its Canals,* Macmillan, New York

Lowell, Percival (1909) *Mars as the Abode of Life,* Macmillan, New York

McCloud, Scott (1994) *Understanding Comics,* HarperCollins

Mark, Kathleen (1995) *Meteorite Craters,* University of Arizona Press

Miller, Walter M. Jr. (1953) "Crucifixus Etiam," in *The Best of Walter M. Miller Jr.,* Gollancz, 2000

Moore, Alan and Gibbons, Dave (1986) *Watchmen,* Titan Books

Moore, Patrick (1998) *Patrick Moore on Mars,* Cassell

Murray, Bruce (1989) *Journey into Space: The First Three Decades of Space Exploration,* W. W. Norton—a wonderful insider's account

Niven, Larry (1999) *Rainbow Mars,* Tor Books

Oberg, James (1981) *New Earths: Restructuring Earth and Other Planets,* Stackpole Books, Harrisburg, PA

Oreskes, Naomi (1999) *The Rejection of Continental Drift: Theory and Method in American Earth Science,* Oxford University Press

Pohl, Frederik (1976) *Man Plus,* Victor Gollancz

Pollack, James (1979) "Climatic Change on the Terrestrial Planets" *Icarus 37,* 479–553

Poundstone, William (1999) *Carl Sagan: A Life in the Cosmos,* Henry Holt

Pyne, Stephen (1980) *Grove Karl Gilbert: A Great Engine of Research,* University of Texas Press, Austin

Raeburn, Paul (with Matt Golombek) (1998) *Uncovering the Secrets of the Red Planet Mars,* National Geographic Society—don't let the lovely illustrations blind you to the excellent text

Relph, Edward (1976) *Place and Placelessness,* Pion, London
Robinson, Kim Stanley (1984) *The Wild Shore,* Futura/Macdonald, London
Robinson, Kim Stanley (1985) *The Memory of Whiteness,* Futura/Macdonald, London—(revision of unpublished *The Grand Tour*)
Robinson, Kim Stanley (1990) *Pacific Edge,* Tor Books
Robinson, Kim Stanley (1992) *Red Mars,* HarperCollins
Robinson, Kim Stanley (1994) *Green Mars,* HarperCollins
Robinson, Kim Stanley (1996) *Blue Mars,* HarperCollins
Robinson, Kim Stanley (2000) *The Martians,* HarperCollins
Schama, Simon (1996) *Landscape and Memory,* HarperCollins
Schivelbusch, Wolfgang (1987) *The Railway Journey: The Industrialization of Time and Space in the Nineteenth Century,* University of California Press
Scott, David H., and Kenneth L.Tanaka (1986) Geologic Map of the Western Equatorial Region of Mars, USGS (I-1802-A)
Seed, David (1996) "The Mars Trilogy, an Interview," *Foundation* 68, Autumn 1996
Sheehan, William (1996) *The Planet Mars: A History of Observation and Discovery,* University of Arizona Press
Shoemaker, Carolyn (1999) "Ups and Downs in Planetary Science" in *Annual Review of Earth and Planetary Science,* 27, 1–17
Smith, Dean (1996) *The Meteor Crater Story,* 1996, Meteor Crater Enterprises
Tanaka, Kenneth L., and David H. Scott (1987) Geologic Map of the Polar Regions of Mars, USGS (I-1802-C)
Taylor, Michael Ray (1999) *Dark Life: Martian Nanobacteria, Rock-Eating Cave Bugs, and Other Extreme Organisms of Inner Earth and Outer Space,* Scribner
Turner, Frederick (1988) *Genesis: An Epic Poem,* Saybrook, Dallas
Watson, Ian (1977) *The Martian Inca,* Victor Gollancz
Westfahl, Gary (2000) "Reading Mars: Changing Images of Mars in Twentieth-Century Science Fiction" in *The New York Review of Science Fiction* 148, December 2000
Wilhelms, Don E. (1993) *To a Rocky Moon: A Geologist's History of Lunar Exploration,* University of Arizona Press

Zelazny, Roger (1963) "A Rose for Ecclesiastes," in *The Doors of His Face, The Lamps of His Mouth and Other Stories,* Faber and Faber

Zubrin, Robert (2001) *First Landing*, Penguin/Putnam

Zubrin, Robert and Maggie Zubrin (eds.) (1999) *Proceedings of the Founding Convention of the Mars Society,* Univelt, San Diego

Zubrin, Robert with Richard Wagner (1996) *The Case for Mars: The Plan to Settle the Red Planet and Why We Must,* Touchstone

Index

INDEX

THE EASTERN HEM

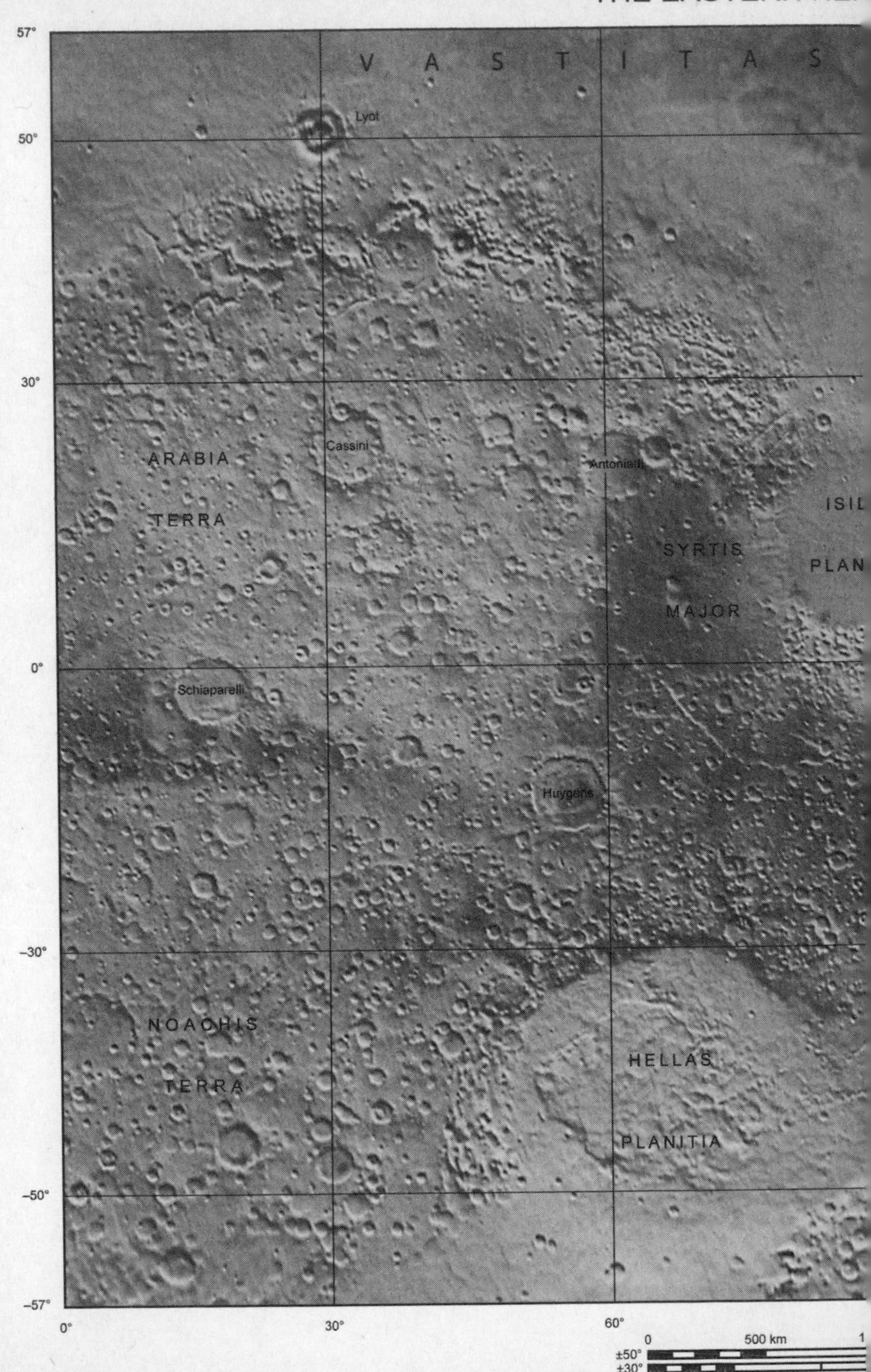

SPHERE OF MARS

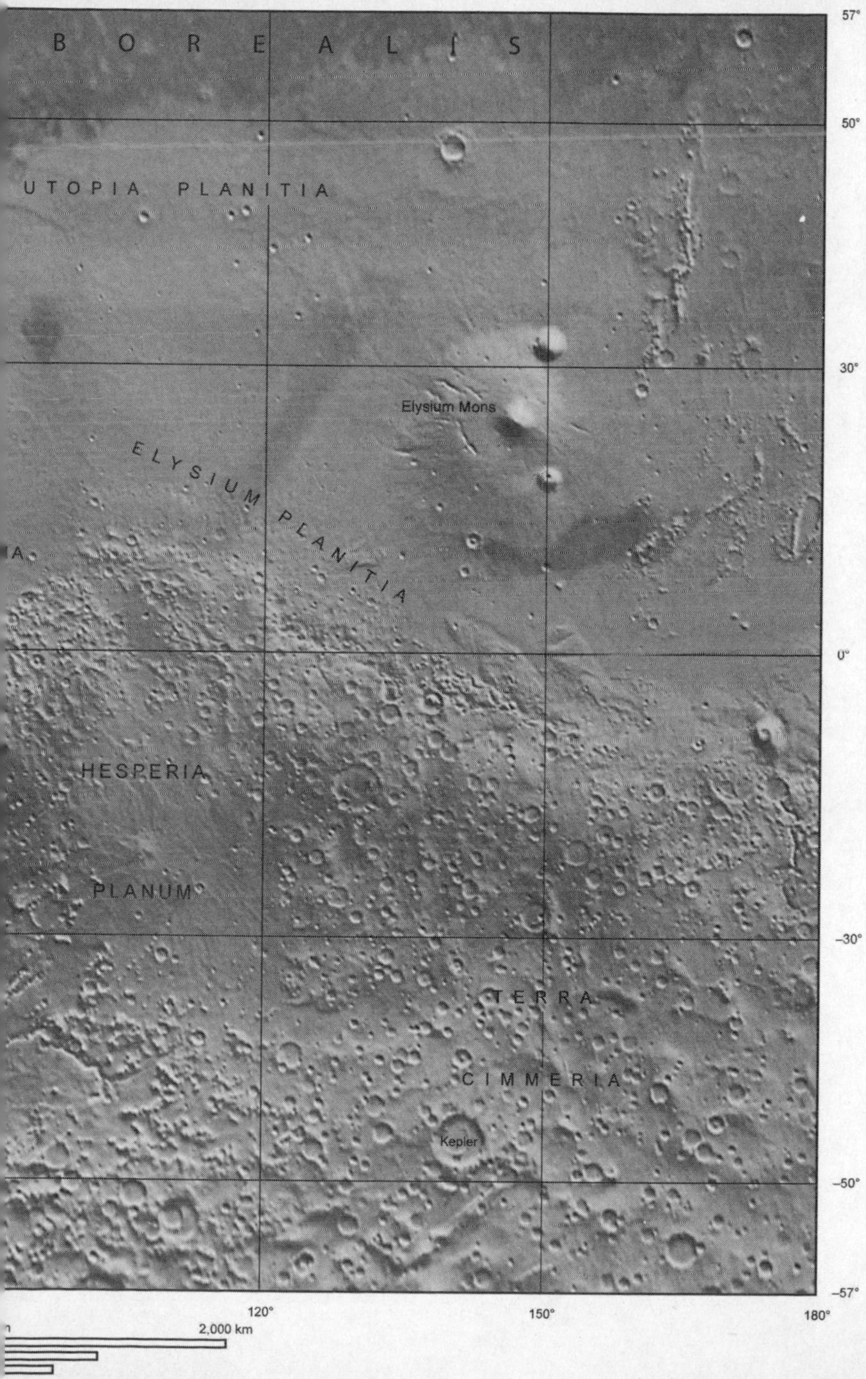